"This is an excellent book. It's a great introduction to the field, and a guide for getting started. It'll be required reading for my MBA students."

Hunter Lovins, President
Natural Capitalism, Inc.
Time Magazine's 2000 Hero of the Planet

"Engineers wanting to lead or serve clients' sustainability interests will find this Guide offers principles as well as useful processes and practices. It is focused and relevant and will help us add value to our clients' projects."

David D. Kennedy, President
Kennedy/Jenks Consultants
San Francisco, CA

"Finally, a book on sustainable development that not only defines the topic, its basis, and opportunities, but provides practical guidance for the successful integration of this challenging market into the consulting engineering community."

Richard A. Millet, P.E., Vice President
URS Corporation
Denver, Colorado

"Bill Wallace provides a comprehensive summary of how the concept of sustainable development is impacting major corporations and other organizations. This book is a must for consulting engineering firms that are planning to serve these potential in the future."

Don Roberts, Former Vice-President
World Federation of Engineering Organizations
(WFEO)
Englewood, Colorado

"Wallace's work on sustainable development is an intriguing exploration of our future's problems and solutions. Insightful to anyone who reads it, it represents a "must have" reference for professionals in the field, as well as a "must read" book for leaders at all levels."

Jeremiah D. Jackson, Ph D., Principal Engineer
Project Resources Inc.
San Diego, CA

"In the past ten years, many of my fellow engineers and I have struggled to understand what sustainable development really means to us as individual practitioners. We readily grasped the concept (and embraced it), but we had difficulty applying it to our everyday world. With this excellent book, Bill Wallace supplies the context for us. Using this book as a guide, we can readily incorporate sustainability into our design ethic – as well we should."

Lawrence H. Roth, PE, GE, F.ASCE
Deputy Executive Director
American Society of Civil Engineers

"Bill Wallace...has developed a special resource – a book which looks at sustainable development from an engineering company perspective. In it, he argues that adopting the precepts of sustainability is not only the right thing to do, but offers a whole new set of opportunities for consulting engineers. The book is a "how to" manual, focused on helping engineering company owners, managers and thought leaders evaluate what sustainable development means to their businesses. It gives them a roadmap for becoming a sustainable company, and identifies new opportunities in the application of sustainable development principles to their engineering projects."

"On behalf of the volunteer leadership of ACEC, I commend Bill for both challenging us all and for giving us considerable resources to reconsider our business models and set ourselves on a sensible and profitable course."

Eric L. Flicker, PE, F.ACEC
ACEC Chairman- 2003-2004

"All engineers should read this thought-provoking challenge to the profession. The questions posed in Becoming Part of the Solution will require us all to provide an honest assessment of ourselves and our profession's commitment to sustainability."

Catherine A. Leslie, P.E.,Project Manager, Associate
Tetra Tech RMC, Inc.
Executive Director, Engineers Without Borders – USA

"Bill Wallace has provided THE reference manual for sustainable development. This book goes beyond the LEED Green Building Rating System and helps the engineering profession become a leader in sustainable development by thinking out of the box. The real life examples that Bill provides are excellent examples of what we are capable of doing as an engineering profession. This is a must read for all new engineers as this will be their future. It is also a must read for all practicing engineers as sustainable development will change how we do business in the future."

David R. Stewart, PhD, PE, President and CEO
Stewart Environmental Consultants, Inc.

"Integrating sustainability and sustainable development in engineering practice is no longer an option for the engineering profession, it is an obligation toward our planet and its people. Bill Wallace's book is very well written and provides a comprehensive look at sustainable development from an engineer's point of view. I plan to use the book in my engineering class on sustainable development next semester. A must read for all engineers and non engineers interested in sustainable development from theory to practice. "

Bernard Amadei, PhD, Professor of Civil Engineering
University of Colorado

"Sustainability issues and trends will mold our future practice as consulting engineers and scientists…like it or not. Becoming Part of the Solution clearly defines the topic and identifies the relevance to firms of all sizes. It also provides an invaluable roadmap to guide our planning for the future survival and success of our firms. We owe Bill Wallace our gratitude for gathering and digesting mountains of sustainability material, and sharing it with us in a clear manner relevant to the engineering practice. The book offers the fundamental guidance to enable us to be leaders in a changed practice environment, and not become its victims as mere service providers."

George F Jamison, PE, Vice President
Goodpaster – Jamison, Incorporated

"Bill Wallace has written an accessible and coherent manual for engineering firms interested in pursuing sustainable development as a new line of business. The manual is a blueprint for firms willing to look beyond the short term and embrace new long-term business ventures that are based on ever-clearer customer and societal needs. Bill has the experience and technical knowledge needed to present the complexity of sustainability as a logical and profitable business opportunity."

Andrew Mangan, Executive Director
United States Business Council for Sustainable
Development

"Bill Wallace's work will provide the framework and I believe the motivation for consulting engineering firms to evolve and market engineering services for sustainable development. I am an engineer and a businessman. I found Bill's arguments that sustainable development was not only necessary for an ever-expanding population but potentially nicely profitable for companies developing the practice were very compelling. Not the least of importance, the book is clear, well argued and very well written."

Charles G. McConnell
President, MAPCO Gas Products Co., Retired

ABOUT THE AUTHOR

Bill Wallace is the Founder and President of Wallace Futures Group, LLC, an organization through which he provides consulting services in the areas of strategic planning, future studies and environmental management. He recently retired from CH2M HILL where, for more than twenty years, he served in a number of senior positions in hazardous waste management, strategic planning, marketing, and new ventures development. He was also CH2M HILL's Liaison Delegate to the World Business Council for Sustainable Development. Currently Bill is chairman of International Federation of Consulting Engineers (FIDIC) Sustainable Development Task Force. He is also Vice-chairman of Engineers Without Borders–USA, a non-profit organization dedicated to improving the lives of people in developing countries through the application of appropriate and sustainable technologies. Bill received a B.S. in Chemical Engineering from Clarkson University, and an M.S. in Management from Rensselaer Polytechnic Institute. He also completed the Advanced Management Program (AMP 104) at the Harvard Business School.

American Council of Engineering Companies
1015 15th Street, NW
Washington, DC 20005-2605
(202) 347-7474

ISBN: 0-910090-37-8

Book design by Sheila Mahoutchian

Printed on recycled paper.

BECOMING PART OF THE SOLUTION

THE ENGINEER'S GUIDE TO SUSTAINABLE DEVELOPMENT

BILL WALLACE

ACEC

AMERICAN COUNCIL OF ENGINEERING COMPANIES

Washington, D.C.

TABLE OF CONTENTS

FOREWORD

Most opportunities for consulting engineering firms have developed incrementally, as linear extensions of what they did today, yesterday, and the week before. Accordingly, we have designed our business models around incremental change, purposely reacting slowly and skeptically to new trends or emerging issues. Thus it is not surprising that the issue of sustainable development—indeed the entire concept of "greening" the built environment—has been dismissed by many of our colleagues as nothing more than a passing fad. Worse, some consider the movement toward sustainable development as harmful to the engineering business.

Bill Wallace is looking to change all that. He has developed a special resource—a book that looks at sustainable development from an engineering company perspective. In it, he argues that adopting the precepts of sustainability is not only the right thing to do, but offers a whole new set of opportunities for consulting engineers. The book is a "how-to" manual, focused on helping engineering company owners, managers, and thought leaders evaluate what sustainable development means to their businesses. It gives them a road map for becoming a sustainable company and identifies new opportunities in the application of sustainable development principles to their engineering projects.

Bill's book is loaded with information, processes, and checklists for investigating, assessing, and deciding how to become a sustainable company. The book also contains references and Web connections to key organizations, documents, and relevant software, providing the user with a tool kit for starting a practice in sustainability. For a company investigating sustainable development issues and opportunities, the book can save literally thousands of hours of research and analysis.

Bill Wallace has a unique background, having served for over 20 years identifying key trends and emerging issues, assessing their impacts, and converting them to new business opportunities for engineering leader CH2M HILL. During this time, Bill served the American Council of Engineering Companies (ACEC) in a variety of roles, including the presidency of the Hazardous Waste Action Coalition (HWAC) and chair of the Sustainable Development Task Force in the International Federation of Consulting Engineers (FIDIC).

Bill doesn't just present facts for our consideration; he challenges us to recognize that our industry may be uniquely positioned to lead—not follow—the movement toward sustainable development. He describes the delicate balance between client-driven and consultant led. Many of us have discussed the need for our client relationships to be up front—understanding the clients' needs and desires before they do. Sustainable development may be the epitome of such an advance opportunity.

I got to know Bill personally as we were serving HWAC some 15 years ago. We saw some things there that were trendsetting that have now become commonplace. Manufacturers were looking at eliminating waste production rather than treating it at the "end of the pipe." In fact, they said that it was cheaper to do so, and the precursor to "green" was born. We often mentioned that green was the color of money—at least in the United States. This mindset has led to selected alternatives that are much beyond the early regulatory drivers.

The book develops a compelling case for incorporating sustainable development into our practices—ranging from the pure social issues, through the direct impacts, to the bottom line. We are challenged to think about the benefits of thinking holistically, for the benefit of both our clients and our firms. Bill addresses the barriers to innovation and proposes a new approach to project delivery, one that develops more-effective partnerships with our clients. Thoroughly sprinkled with specific examples of real-life applications, ranging from corporate policies to municipal transit projects, the book provides an interesting and readable treatise on why adopting the principles of sustainability makes business sense.

It is clear that our various client groups will have different views of the impacts of sustainable development. Some may consider it an impediment to their operations; others consider the application of sustainable development policies a key ingredient for their long-term business success. Bill's book offers ideas on how to approach our clients. We must be careful not to preach sustainability, but rather offer it up as an emerging concept pulled from our firm's experience working across many client sectors. Bill points out that following sustainable practices makes business sense. Projects in which sustainable development principles are applied comprehensively have a lower life cycle cost and are more acceptable to the public. Furthermore, companies see sustainable development as a catalyst for innovation, developing new products and services based on principles of eco-efficiency and eco-effectiveness.

The challenges are great, but so are the opportunities. Over many years, ACEC has clearly identified many specific business-related issues, and it has done a marvelous job at representing our constituency. Sustainable development may not be as easily identifiable, but surely it will have many more far-reaching implications affecting our practices for many years to come. It is essential for all of us to take the lead in making our future development sustainable. This role is not new for us. In the past, we led the development of safe drinking water, sanitary waste disposal, and the myriad things improving our quality of life that engineering has made possible and that the developed world now takes for granted.

On behalf of the volunteer leadership of ACEC, I commend Bill for both challenging us all and for giving us considerable resources to reconsider our business models and set ourselves on a sensible and profitable course.

Eric L. Flicker, PE, F.ACEC
ACEC Chair, 2003–2004

PREFACE

Humanity has the ability to make development sustainable—to ensure that it meets the needs of the present without compromising the ability of future generations to meet their own needs. The concept does imply limits—not absolute limits but limitations imposed by the present state of technology and social organization on environmental resources and by the ability of the biosphere to absorb the effects of human activities.[1]

It is very clear, from our perspective, that to create a successful business we have to embrace sustainable development.[2]

For us at DuPont, the question of whether sustainability improves the bottom line is not asked any more. We are committed to sustainable growth. That is how we will get our bottom line, and that is how we will create value for our shareholders and for society.[3]

Purpose and objectives

My purpose in writing this book is to open the door for engineers onto a whole new line of business: engineering services for sustainable development. I wanted to cut through the current fog of ideologies and agendas and present a clear and compelling story of why sustainable development will be an important issue and a new way of thinking for engineers and their clients, indeed for everyone, in the decades to come. I intend to show the readers that sustainable development represents a new and exciting business, one that is broad in both scope and opportunity for engineering firms of any size or scope of business.

What does sustainable development mean? How is it different from conventional economic development? What does it mean for engineers? The term *sustainable development* was popularized in the 1987 United Nations report *Our Common Future*.[4] This landmark report showed that, over the last century, our collective ability to produce growth and prosperity has also had significant unintended effects on our resources and ecological systems. These effects appear to be growing so large and so negative that they may compromise the ability of future generations to meet their needs. To reverse these trends, the report called for a new form of economic development—sustainable development—that would enable society to meet today's needs while conserving resources and natural systems for future generations.

Sustainable development is a goal, not an ideology. Making our development sustainable means shifting away from our heavy use of nonrenewable materials and energy resources. It means creating new materials and inventing technologies that make better and more-efficient use of renewable and nonrenewable resources. It means restoring and preserving our renewable resources and ecological carrying capacity, making them more productive to accommodate an expanding and increasingly demanding population. Finally, it means the fair and equitable use of these resources, among populations, nations, and generations. Sustainable development seeks to strike a balance among three elements: economic growth, environmental protection, and social equity.

For engineers, the journey toward sustainable development opens up a whole new set of challenges and opportunities. To reach the goals of sustainable development, engineers must find new ways to meet society's resource and energy needs using renewable supplies. They must reduce substantially the damage and burden to our ecological systems by reducing (or better yet, eliminating) by-product emissions to the air, water, and land. They must restore the abundance and productivity of our natural resources.

Moving toward sustainability will require more or less a complete overhaul of the world's infrastructure, replacing or refurbishing existing systems with new, cleaner, and more-efficient processes, systems, and technologies. As such, new world markets for sustainable engineering services are being created as industries and governments alike begin the changeover to more sustainable practices. Providing engineering services for this changeover will require new forms and levels of engineering experience and expertise, much of which is currently unavailable. Nimble companies that can move quickly into these markets will find that they can price their services at levels substantially above the norm.

Warning signs

The symptoms of nonsustainable development are starting to appear in many places and at many scales. In the United States as well as in other developed countries, cities are now feeling the effects of urban sprawl, as people relocate to suburban areas seeking a better quality of life. Cities are left with an eroding infrastructure and a diminishing tax base while the roads and highways leading into the city become more congested. Urban runoff, the application of agricultural chemicals, and other mechanisms have polluted lakes, rivers, and estuaries, damaging habitats and destroying fishing resources. Contaminants are accumulating in increasing quantities in the receiving waters. A huge "dead zone" now appears annually in the Gulf of Mexico, covering an area of more than 5,000 square miles. The recent drought and resulting water shortages in the western United States demonstrate the precariousness of our water resource base.

On a larger scale, there is the matter of global climate change. Although still controversial, the consensus view is that many of the gases produced primarily through the combustion of fossil fuels are causing an enhanced greenhouse effect that will change climate patterns significantly in this century. Individually, these examples appear to be isolated and perhaps addressable. Aggregated together along with hundreds of other examples, the evidence shows a collective degradation of the environment on all scales.

To the people who live in the developed world, sustainable development may appear to be just another cause in a succession of environmental causes, many of which are worrisome but for which nothing much can be done. However, a growing number of companies and public sector organizations (many of your clients) have a different view. Public concerns over global climate change, nonpoint source pollution, water shortages, deforestation, loss of biodiversity, environmental justice, and other problems have prompted companies and public agencies to change the way they produce and deliver goods and services to their customers. Many have responded by making public commitments to reduce their so-called ecological footprint, a measure of environmental impact of their operations. Companies and other organizations have also taken on social responsibilities for education and welfare, responsibilities that were largely the dominion of governments. Rather, they are not necessarily altruistic. Their motives are born out of the recognition that their key stakeholders—employees, stockholders, customers, affected communities, and others—are seeing the effects of nonsustainable behavior and holding the responsible organizations to the stakeholders' own higher standard of performance. These companies recognize that in today's "CNN world," people, in effect, give organizations the license to operate, and people can revoke that license at any time.

The "take-make-waste" production-consumption model

Our quality of life is derived from a linear, "take-make-waste" production-consumption model of the industrial age, one that draws freely upon energy and raw materials to produce and deliver goods and services to meet consumer needs. As a result, Americans consume more resources, in terms of both absolute and per capita consumption, than any other country in the world. This model is based on the underlying assumption that the earth's supply of resources and ecological carrying capacity (its ability to assimilate wastes and ecological services) is more than sufficient to support all the activities of an ever-increasing and demanding population.

Underpinning our current, nonsustainable system is a huge array of legacy operations, processes, and technologies that were designed using old assumptions about the use of resources and effects on ecological carrying capacity. We have always assumed (and history supports) that resources will always be plentiful and that ecological damage is always fixable or short lived. If a resource shortage appears, then we figure out how to find more, or we create substitutes. If we pollute a river or a lake, we usually find a way to clean it up. This strategy has always worked in the past. However, things are about to change. World population growth coupled with economic expansion are placing unprecedented demands on resources and ecological systems, demands that cannot be met by patching up mistakes or waiting for technology to save the day. Sustainable development means doing more with less—inventing technologies and systems that offer the same or better services, but at a lower resource, ecological, and social cost.

It will take many decades of sustainable technology invention and deployment before this system can be made sustainable. Industry and government are just beginning to make these changes, changes that will be accomplished incrementally, project by project, mostly by engineers. To these ends, this book offers methodologies for delivering sustainable projects and measuring the contribution of those projects to sustainability. It also discusses the policies and mechanisms for sharing knowledge and best practices across industry and government.

This world makeover will require a new array of technologies and processes, most of which have yet to be invented. Here engineers play a key role. They will help determine not only which technologies and processes are appropriate under a sustainability schema, but if and how they can work together reliably and effectively to produce the desired result. Firms wanting to participate in this work will have to assemble the requisite resources and skills, many of which are new and still evolving. These new tools and disciplines include life cycle analysis, corporate social responsibility, green chemistry, carbon emissions reporting, carbon trading, industrial ecology, and eco-efficiency, to name a few.

Engaging engineering firms in sustainable development

Convincing your firm that it ought to recognize and participate in this new area of business may not be easy, particularly for firms whose clients are based largely in the United States. With the exception of multinational corporations, most of the leaders and managers in U.S. public- and private-sector organizations have not experienced the resource shortages or ecological problems associated with nonsustainable activities. The United States still enjoys an abundance of resources that, when combined with its technological ingenuity, tends to obscure problems associated with sustainability. In addition, the United States has amassed over many years a comprehensive set of environmental laws, regulations, and programs, which arguably deal with a number of the concerns raised by advocates of sustainable development. Thus, for many, it is hard to see sustainable development as an important issue and, subsequently, as a problem deserving attention.

Here again, changes are taking place. The forces of globalization are altering the way we in the United States see the world and how the world sees us. Stakeholders are demanding better environmental stewardship as well as broader acceptance of social responsibilities. What is more, they are getting it. Economist Milton Friedman's old axiom, "The business of business is business" is no longer true. As a result of market forces and public opinion, corporate responsibilities are extending up and down the supply chain. Companies that use wood in their operations or sell wood products are now seeking wood sources from sustainably managed forests. Manufacturers are setting high standards for working conditions in their own and their suppliers' workplaces across the world. Buildings and factories are being designed or refurbished with drastically reduced environmental footprints, that is, lower energy usage, use of recycled or renewable materials, extensive use of daylighting, energy obtained from renewable sources, and more. Most important, these changes are reducing operating costs and creating higher profits.

In addition, many companies view the movement toward sustainability as an opening to an entirely new set of market opportunities. They are devising new products and services to meet newly discovered customer needs. For example, developers and building owners are creating so-called green buildings and facilities. They not only cost less to build and operate, but offer better and more-productive environments for the occupants. Power companies are selling green energy at a premium to customers who want to contribute to reducing their use of nonrenewable energy resources. Chemical companies are producing new industrial-grade plastics made from renewable resources. Automobile manufacturers are introducing hybrid vehicles that have reached record levels of fuel efficiency. Today buyers of these vehicles must get on a waiting list and wait as long as 12 to 18 months until their order arrives.[5]

Correspondingly, new markets for sustainable development services are opening up to engineering firms to help clients improve their performance, save money, and become more sustainable in the eyes of their stakeholders. These opportunities are available to all engineering firms, regardless of size or client base. In fact, it can be argued that in this market, the small- to midsized firms have a distinct advantage over large engineering firms. Why? For one thing, the small- to mid-sized firms usually possess a better environment for innovation, one that is more collaborative, nimble, and amenable to change. Furthermore, the firm's principals tend to have closer relations to their clients, and thus are better able to identify and aggregate the sustainability trends and market forces that affect their clients. Finally, because they are small, the leaders of these firms can make changes more quickly than their larger counterparts. That is because they have closer relationships to the managers and thought leaders of the firm, and because they have more control of the operation.

However, despite these significant opportunities and advantages, these business opportunities have slipped past the engineering community. For the most part, its members either have not analyzed the trends sufficiently or have dismissed them as being faddish or even antibusiness.

Lack of an environment for innovation

Why has the engineering community missed these opportunities? Most of the people who I have spoken with tell the same story. Their firms, particularly the larger firms, do not foster an environment for innovation. I define this environment as a creative workplace that encourages individuals to systematically search for new markets and technologies, share knowledge across organizational boundaries, and quickly pool or acquire the resources necessary to pursue new areas of business.

There is a sad irony here. Engineering firms are perhaps best positioned to reap the advantages of this sort of innovative business environment, but have one of the poorest track records for doing so. These firms generally work closely with many clients in many locations and in many sectors of the economy. Thus they enjoy a unique vantage point, a special window on the trends, issues, and forces at work in multiple industries. If they do things right, they can become deeply involved and astonishingly knowledgeable about their clients' businesses. Armed with this great knowledge and staffed with bright people, they are better equipped than any other group for identifying and interpreting new trends and threading together new opportunities across a wide business and economic landscape.

Instead, firms tend to stifle innovation by setting up reward and recognition systems that, in reality, reward revenue growth and profitability first and foremost. Despite the rhetoric, all other measures fall to a distant second. Clearly, revenue growth and profitability are important objectives. However, they are short-term measures that do not reflect the long-term health of the firm, nor do they reflect how well a firm is leveraging its knowledge and resources for long-term success.

The result is organizational "stovepiping," often done in the name of cost efficiency. In this environment, business units become insular, focusing their energies and resources within their own domain. Sharing information across business unit boundaries is an action against their best interests. Without some way of recognizing the benefits of cross-business-unit cooperation, reward and recognition becomes a zero sum game, with each unit seeking to improve its own financial performance at the expense of the others.

Delivering engineering services in sustainable development

Success in delivering sustainability services to clients must be based on a sound business case, that is, following the precepts of sustainability adds to shareholder value. This book shows how to present such a case to clients using case examples of successful projects and the benefits derived therefrom. One key to success is being able to "walk the talk," that is, demonstrate to clients that the engineering firm also sees the business and ethical value in sustainability, and that its leaders, managers, and staff have made a commitment to sustainable development.

This book will enable engineers and managers at all levels to develop for their own organizations a business case for sustainable development. It will help them answer the questions of why and how much they should invest in the resources and tools necessary to create a sustainable development practice. The book also shows how leaders and managers of engineering firms should prepare their organizations to engage in this new business opportunity.

Finally, the book looks to the future, raising and addressing some of the critical issues of sustainable development for the next several decades. What are the critical resource and ecological carrying capacity issues? What technologies are needed to effect the necessary changes? What do government and industry need to do? How should the engineering community respond?

Organization of this book

As noted at the beginning of this section, this book is intended to provide the reader with not only a working understanding of sustainable development, but also a practical guide for starting up a sustainability practice and delivering projects that contain elements of sustainable development. To these ends, this book is organized into seven chapters plus three appendices.

- *Chapter 1: Introduction.* Provides a definition of sustainable development and the history of its development.

- *Chapter 2: Sustainable Development: Origins, Concepts, and Principles.* Discusses the basic principles of sustainable development to give the reader a solid grounding in the principles and related issues.

- *Chapter 3: Sustainable Development: Client Needs and Market Drivers*. Discusses the trends and market forces that are creating the market for sustainable development services.

- *Chapter 4: Walking the Talk*. Describes the benefits of making an organizational commitment to sustainable development and how an engineering firm should consider making such a commitment.

- *Chapter 5: Greening the Engineering Company*. Presents programs for setting policies, procedures, and practices for reducing a firm's environmental footprint.

- *Chapter 6: Delivering a Sustainable Project*. Describes how to set up and successfully conduct a sustainable development project.

- *Chapter 7: The Future of Sustainable Development*. Considers how the issue of sustainable development will evolve over the next five to 10 years and what that might mean to engineering firms.

- Appendices:

  - Appendix A: Useful Sustainable Development Publications
  - Appendix B: Sustainable Development Resources and Tools (on CD-ROM)
  - Appendix C: Sustainable Development Principles

Becoming part of the solution

The engineering community is truly at a crossroad. We can continue what we have done, which is simply to support the current production-consumption (take-make-waste) model. We can continue to find ways to extract more resources faster and cheaper than before. In the face of an ever-increasing consumer demand, we can make, transport, and sell more goods to more people everywhere. We can help our clients do all of this while staying under regulated limits of pollution emissions, cleaning up what slips through. In this role, engineers are part of the problem, finding ways to perpetuate an illogical system and calling it progress. It is commodity engineering work characterized by slim margins and intense competition, increasingly from abroad.

The purpose of this book is to show a better way, a way for the engineering community to become part of the solution. Instead of finding ways to extract resources faster, we can be inventing and applying new technologies that use less materials and energy. Instead of finding ways to sell more products, we help clients get more service per unit of product. We can find ways to use natural systems to serve our needs for lighting, heating, and cooling. We can design buildings and other structures for flexibility in use, reuse, and recyclability, thereby reducing life cycle costs.

Pursuing this course will bring about new engineering challenges, challenges that will force us to work smarter and call upon a broad set of skills and resources. These are the sorts of challenges that can attract young people into engineering, showing them how they can apply what they learn to make a difference in the world instead of following old and discouraging pathways.

Acknowledgments

I wish to thank the American Council of Engineering Companies (ACEC) and the members of ACEC Environmental Business Committee for giving me the opportunity and the encouragement to write this book. Eric Flicker, Mary Goodpaster, Jerry Jackson, George Jamison, David Kennedy, and David Stewart were especially helpful in providing me with their thoughts and comments throughout the writing process.

Much of the credit for connecting the issues of sustainable development to the work of engineers should go to Don Roberts, who first alerted the engineering community to this important issue, and then persevered to help them understand it. Additional credit should go to Professor Bernard Amadei, of the University of Colorado at Boulder. Through his pioneering work with Engineers Without Borders-USA, Professor Amadei is showing the engineering community how its knowledge and talents can truly contribute toward the making of a better world.

To the members and staff of the World Business Council for Sustainable Development and its U.S. regional counterpart, I want to express my appreciation for their continuing efforts to communicate the importance and benefits of incorporating sustainability into everyday business practices. My role and participation as CH2M HILL's liaison delegate to the council provided me with important grounding on the issues of sustainability. It also enabled me to work with many exceptional people and see the issues of sustainable development from a global perspective. Thanks also to some of the exceptional people at CH2M HILL–Jo Danko, Jan Dell, Scott Johnson, Linda Morse and Andrea Ramage–for sharing their thoughts and materials on sustainable development projects and corporate greening.

Special thanks to Jerry Plunkett for many years of mentorship and encouragement and for showing me how to really think outside the box. Also, my sincere gratitude to Bob Dempsey for showing me that there was a box to think outside of.

Last but certainly not least, to my wife, Diane, not only for her excellent editing and suggestions, but most of all, for her support and constancy through this project and in all things. To our daughters, Kate and Jane, who inspired this work, my apologies. It seems my generation has gotten your world into a bit of a mess.

Bill Wallace

Steamboat Springs, Colorado

November 23, 2004

---

[1] United Nations, World Commission on Environment and Development, *Our Common Future* (New York: Oxford University Press, 1987), 8.

[2] Carlos Guimaraes, vice president of Dow Chemical Corporation and chairman of the U.S. Business Council for Sustainable Development, quoted in Business Week Online, May 1, 2003.

[3] Chad Holliday, CEO of DuPont, quoted in Charles O. Holliday Jr., Stephan Schmidheiny, and Philip Watts, *Walking the Talk: The Business Case for Sustainable Development* (San Francisco: Greenleaf Publishing, 2002), 24.

[4] World Commission on Environment and Development, *Our Common Future*.

[5] Amanda Leigh Haag, "Boulder Demands Hybrid Vehicles: Civic Hybrid, Prius Sell Faster than They're Made," Boulder Daily Camera, May 29, 2004.

CHAPTER 1

Introduction

Sustainability represents the number one megatrend that will change the rules of global business. ... Innovative companies will surely find ways of exploiting logic for gaining relative advantage via lowering costs, differentiating products and service, leveraging experience and core competencies and discovering new market and geographic strategies.[1]

Background

Just a 10-minute walk from my former home in Littleton, Colorado, sits a light-rail station. Opened in July 2000, it is the first point of entry to a 15-mile ride along the southwest corridor to Denver's downtown business, sports, and recreational centers. Before it opened, critics declared that the system would see limited use and was therefore a waste of taxpayer money. But on opening day, commuters, fed up with fighting heavy rush hour traffic, overflowed the station parking lot and filled the trains to downtown Denver—a commuting routine that continues to this day. Parking capacity the transportation planners deemed adequate for many years was filled on opening day, forcing commuter parking to spill over to adjacent streets and properties. The number of parking spaces was quickly doubled, but parking capacity still falls short of demand. Today more light-rail lines are being planned for the metro-Denver area, and the critics of Denver light-rail are uncharacteristically silent.

Figure 1-1: Light-rail station in Littleton, Colorado. Passengers debark from a light-rail train that has just arrived at the Mineral Avenue Station. Photo courtesy of Dr. Jon Bell, Presbyterian College, Clinton SC.

In addition to providing transportation downtown, the Denver light-rail is changing the landscape along its path. Stations built along this southwest route are stimulating urban renovation and development in a way that highways cannot. Within walking distance of each of the Denver light-rail stations, new shopping centers, residences, and commercial buildings are being built to accommodate the shifts in commuter movement and market demand, reversing to some extent urban sprawl.

Denver is not alone in its light-rail success. After opening in 1999, Salt Lake City's 15-mile light-rail line is now carrying an average of 20,000 people on weekdays—43 percent above projections. Dallas's ridership is 30 percent above the city's estimates, while St. Louis's is 14 percent above plan.

What is going on here? Why were the estimates so consistently low? Supporters of light-rail point to out-of-date planning models and prescriptive estimating rules handed down by the Federal Transit Administration (FTA). These rules do not take into account new public attitudes and values. The FTA models assume that when it comes to a choice of bus or rail, the public is indifferent. In reality, the public favors rail over bus transportation by a wide margin. A Denver Regional Transportation District survey found that 60 percent of the people who had never ridden the bus now use the light-rail three times or more a week.[2]

Light-rail substantially reduces traffic congestion, which translates to a better quality of life. Having this transportation option means better access to downtown businesses and reduces the need to move outside the city, closer to where people live. Furthermore, people are getting tired of urban sprawl and are making their wishes known at the voting booth. In 2002, 94 communities in 23 states passed $2.9 billion in smart-growth legislation and other initiatives for open space conservation and restoration. Overall in 2002, voters approved a total of $10 billion in conservation and conservation-related funding.[3]

Figure 1-2: Rush hour traffic in a U.S. metropolitan city. Source: JupiterImages, www.clipart.com.

Light-rail projects and smart-growth initiatives are trends to be watched, not simply for their impact on engineering and construction work but for what they signal as future patterns of change. People are pursuing a better quality of life, defined in part by fewer commuting hassles and more open space. Building more highways does not address their needs.

Renovation for sustainability and profitability

In Dearborn, Michigan, the Ford Motor Company is refurbishing its historic River Rouge production plant in an unusual way. Not only is the plant being designed to be a highly efficient production facility, but it also incorporates cutting-edge technologies that are environmentally benign and energy saving. Project architect and sustainable development guru William "Bill" McDonough fitted this large industrial complex with a 10-acre organic roof planted with a low-maintenance perennial. Responsibilities for design and construction were given to the firms ARCADIS Giffels, LLC and Walbridge Aldinger. The design also incorporates the use of natural lighting (daylighting) and natural storm-water treatment. The River Rouge plant makeover is the inspiration of Ford CEO William Clay Ford Jr., part of his vision for Ford in the 21st century.

Figure 1-3: The new, living roof is one of the prominent features of an innovative $2 billion modernization of Ford's 84-year-old Rouge River manufacturing complex in Detroit, Michigan. © Ford Rouge Design Team. Photo courtesy of William McDonough and Partners.

ARCADIS is incorporating "green" concepts into its design of the River Rouge plant. It is designing the nation's largest porous parking lot, a design intended to capture and direct storm-water runoff to a natural (plant-based) treatment system. ARCADIS has incorporated green design in other projects. Its recent design of a U.S. Postal Service mail distribution facility uses natural lighting, low-energy air-conditioning systems, and solar shading. Walbridge Aldinger, ranked first by the *Engineering News-Record* in automotive construction for 2003, is providing environmentally friendly design and construction services not only to Ford, but to other automotive clients. It is designing Visteon Village, the future corporate headquarters of Visteon Corporation, by reclaiming a gravel quarry in a wooded area outside Detroit, Michigan. It is seeking LEED certification[4] as a facility built in accordance with sound environmental principles.

Creating new materials from renewables and industrial by-products

Ford isn't the only company that is bringing environmental principles into its design. Dow Chemical Company recently established a joint venture with Cargill, Inc. to manufacture industrial-grade plastics from renewable resources. Using their patented process called NatureWorks, corn and other agricultural crops are processed to create materials for a wide range of fiber and packaging applications.[5]

Dow is also starting a project in *by-product synergy*, based on the concept that the unwanted by-products from its facilities should be assessed for their potential as feedstocks for other facilities in the Dow organization. Like other companies in the chemical industry, Dow process engineers know their processes thoroughly and have worked hard to minimize or find uses for all process wastes (unwanted by-products). What they hadn't done, as least up until now, is look beyond their traditional boundaries and see what materials other Dow companies are using for feedstocks and what unwanted by-products they may be generating. Matchups (one company's waste is another company's feedstock) can save the company plenty in terms of reduced materials and waste disposal costs.

LEED Green Building Rating System

The LEED (Leadership in Energy and Environmental Design) Green Building Rating System is a scoring system for rating buildings based on sustainability criteria. It is being developed and administered by the U.S. Green Building Council (USGBC). Buildings are scored on a point scale and can achieve four levels of certification based on the points achieved. Additional information about LEED can be found in chapter 6.

- LEED certified: 26–32 points
- Silver level: 33–38 points
- Gold level: 39–51 points
- Platinum level: 51+ points (out of 69 possible points)

What is by-product synergy?

By-product synergy is the brainchild of Dr. Gordon Forward, retired vice-chairman of Texas Industries, Inc. (TXI). Forward was the president and CEO of Chaparral Steel, a company listed in Robert Levering and Milton Moskowitz's "100 Best Companies to Work for in America." The way Forward tells it, TXI (Chaparral Steel's owner) asked him to manage TXI's cement plant, just down the road from Chaparral Steel. Six weeks after taking over the cement plant, Forward discovered that the steel slag Chaparral was paying to get rid of could be used, after a few chemistry adjustments, as a feed material in the making of high-quality portland cement. Not only would both companies save money, but the resulting cement would be of better quality. Furthermore, incorporating steel slag in the cement-making process resulted in a 10 percent increase in productivity, lowered energy requirements, and reduced greenhouse gas emissions.[6]

"It was amazing," recalls Forward. "I've been driving by the cement plant for 30 years on my way to work, and never imagined that such a business opportunity existed."[7]

TXI later patented the process and named the product CemStar. The process is now licensed around the world. In September 1999, the U.S. Environmental Protection Association (EPA) awarded TXI its international 1999 Climate Protection Award for the CemStar process. Later that same year, the company was awarded the EPA's ClimateWise Award for its efforts in reducing energy consumption in its operations.

TXI was a member of the Gulf of Mexico Business Council for Sustainable Development. Soon after this discovery, Gordon Forward began working with Andy Mangan, then the council's executive director, on how to develop this idea.

"We thought, If two companies can get together and create a product like CemStar, what would happen if we got twenty companies from different sectors working together?" recalls Andy.[8]

In cooperation with the council, Forward and Mangan created a process they termed *by-product synergy*. In this facilitated process, representatives of 15 to 25 companies from different industrial sectors come together and exchange information about their feedstock needs and the by-products generated. After learning who needs what and who produces what, they use computers and detailed data tables to match needs with supplies. For

Figure 1-4: CemStar system installed in a Texas Industries, Inc. cement plant. Photo by Bill Wallace. Permission courtesy of Texas Industries, Inc.

the first project, 21 companies worked together and produced over 60 synergy possibilities. Out of these, 13 synergies were pursued by the participating companies. Additional information on by-product synergy can be found in chapter 2.

Sustainable growth at DuPont

In 2000, DuPont made sustainable growth a "core direction" for the company, one that meets society's needs and values while creating substantial shareholder value. In his 2002 *Sustainable Growth Progress Report*, Charles Holliday Jr., DuPont's CEO, stated: "Our goal for the 21st century is to become a sustainable growth company—one that creates shareholder and societal value while decreasing our environmental footprint along the value chains in which we operate. As part of our transformation we have worked hard on reducing our environmental impacts and have set aggressive targets to be attained by 2010 in the areas of energy use, greenhouse gas reductions and the use of renewable energy and feedstocks." [9]

In a 2003 speech to corporate executives, Dr. Uma Chowdhry noted that DuPont, like other manufacturing companies, has grown by making and selling more and more products. Consequently, its financial growth was tied directly to how much energy and materials it used and wastes it created. That approach was not sustainable over the long term and needed to change. To these ends, DuPont has set four goals for 2010:

- To derive 25 percent of revenues from non-depletable resources–up from 14 percent in 2002.

- To reduce global carbon equivalent greenhouse gas emissions by 65 percent, using 1990 as a base year. The company has already surpassed this goal with a 68 percent reduction.

- To hold energy use flat using 1990 as a base year.

- To source 10 percent of the company's global energy use in the year 2010 from renewable resources. [10]

With sustainable growth as its platform, DuPont has created new products that not only make a profit but have environmental and societal benefits as well. For example, its new cotton insecticide, Avaunt, helps African farmers grow cotton safely and economically by reducing the required dose per hectare. Cotton yields are increased 15–20 percent, and low-dose applications (nearly 10 times less than the older products) help protect the health of the farmer.

In addition, DuPont's new Super Solids technology helps automobile manufacturers reduce volatile organic compound (VOC) emissions by 25 percent in the automotive painting process. This helps its customers, the automotive manufacturers, reduce their toxic emissions while producing a more durable auto paint finish. Furthermore, Scientific Certification Systems certified DuPont's Antron nylon as an environmentally preferable product: a product that has a "lesser or reduced effect on human health and the environment when compared with competing products or services that serve the same purpose."[11] In terms of life cycle analysis, Antron is more durable, consumes less materials and energy in its manufacture, and generates less waste when compared with competing products.

Figure 1-5: DuPont Avaunt insecticide helps cotton farmers in West Africa defeat pests and boost yields, yet has a low impact on the environment. Photo courtesy of DuPont.

STMicroelectronics' ecological vision

Like DuPont, the $8 billion semiconductor manufacturer STMicroelectronics has made environmental responsibility a key component of its corporate strategy. CEO Pasquale Pistorio has made a public commitment to sustainability, believing that such policies and actions significantly enhance competitiveness. By accepting a "moral responsibility" to the environment, Pistorio believes he can not only attract the best people, but reduce significantly the overall cost of operations. Payback to date on environmental investments has been over $1 billion. Its ecological vision is "to become a corporation that closely approaches environmental neutrality. To that end we will not only meet all environmental requirements of those communities in which we operate but, in addition, we will strive to comply to the following ten commandments."[12] These 10 commandments are stated in the STMircoelectronics Environmental Decalogue and presented in chapter 5.

To stay competitive in the semiconductor industry, STMicroelectronics is constantly upgrading its chip fabrication facilities, spending millions of dollars in new construction each year. These facilities are designed and built not only to meet high standards of productivity, but also to meet the company's environmental responsibilities.

Sustainable cities: Seattle and Austin

In Seattle, Washington, city policies require that all city-funded projects and renovations of over 5,000 square feet earn a silver rating using the LEED rating system. Seattle expects to net an annual savings of $500,000 per year on 11 buildings. There is also the added benefit of an improved work environment, which results in lower turnover, increased productivity, and fewer sick days. This is important, given that ninety-two percent of building operating cost is employee payroll. This LEED silver rating requirement and other activities made Seattle the number one sustainable city in the United States, as rated by Kent Portney in his new book, *Taking Sustainable Cities Seriously*.[13] According to Kent, there are 24 cities in the United States with substantial sustainability programs.

Similarly, the city of Austin, Texas, has recognized the benefits of building green. Its award-wining green building program has been in place since the 1980s. Today the program provides extensive guidance and tools for residential and commercial builders on energy and water conservation. In addition, it has established rebate programs for residents and owners of commercial buildings wishing to improve their energy and water conservation performance. Most important, the city has instituted measures to bring green building into the mainstream. It has established green rating programs so that developers, realtors and home buyers can recognize the qualities and benefits of green homes.

Economic benefits of building green

Still, for many corporate decision makers, the prospect of designing a building to be green sounds like an additional expense. However, this is not necessarily true. Scott Johnson of CH2M HILL presents the economic case for green buildings in a series of tiers of economic costs and benefits.[14] Johnson notes that there is potentially great synergy between building green and building cost effectively. It is, however, very important to note that this doesn't just happen on its own. All so-called green building investments do not produce a positive return on investment (ROI). It is important to design the building to be cost effective while concurrently designing the building to be green. It is folly to claim that all green criteria (e.g., the LEED criteria) will translate to an improved bottom line, or that all benefits have an equal level of certainty attached to them.

Figure 1-6: Seattle, Washington, City Hall. A building with many sustainable innovations. The building uses a "raised floor" system in the office areas, which allows utilities to be moved wherever they are needed, increasing flexibility and reducing renovation waste and cost. It also improves energy efficiency and allows for individual occupant control of temperature and airflow. Photo courtesy of Emily Burns of the City of Seattle.

When speaking to building owners, Johnson has a compelling argument for building green. He starts with the easy, lower-level tiers and works his way up, racking up significant savings at each level using the owner's own methods for analyzing the economics. He builds the business case using the types of green cost efficiencies that are likely to be "sellable" to the building owner based on the timing and the certainty of the benefits. Done correctly, such an approach can add millions of dollars to the owner's bottom line.

"The industry's own benchmark for performance is total cost of ownership," Johnson begins.

> *This includes not just the up-front design and construction costs, but the annual operation and maintenance too. Today that's usually three times the cost to build the building. However, too often, decisions are made strictly on first costs.*

Now let's talk about productivity. What about the people that occupy the building? The cost of design-ing, constructing, operating and maintaining the building only amounts to 8 percent of the total cost for the economic utility provided by the building and the people in it. The rest, a whopping 92 percent, is made up of employee salary. There are studies that indicate that you might get upward of 20 percent improvement in productivity out of a building that is designed with open space and daylighting. Don't buy that? What would you pick? 5 percent? 2 percent? What about 1 percent? If you're a knowledge company, a 1 percent improvement in the productivity of your workforce might be worth doubling the cost of the building. But it is unlikely that would be necessary. A small percentage increase in the first costs of the building (or perhaps no increase at all) might provide an ROI of hundreds of percent.

Johnson goes on to discuss the potential contributions to corporate image and ex-ternal factors. These benefits are harder to convert to hard currency, but may also have real value. All in all, he has voiced a strong business case for investing in green build-ings by looking at the issue on a whole life cycle basis and using measures that are not necessarily obvious.[15]

Figure 1-7: Zack Elementary School in the Poudre School District in northern Colorado. This school has won national awards for its conservation and energy-saving features. The District's sustainable school designs cost less to build and operate than conventionally designed schools. Photo courtesy of the Poudre School District.

Do buildings designed using more sustainable technolo-gies cost more to build? Not necessarily, according to Bill Franzen, executive director of operations of the Poudre School District in Fort Collins, Colorado. Franzen's organiza-tion builds schools that are not only cheaper to operate, but cost substantially less to build than conventional school buildings.[16] Furthermore, these buildings provide better lighting, consume less energy, and conserve materials and natural resources.

Unrelated anecdotes or signs of a new emerging issue

Are the previous items just interesting anecdotes, or are they signs of an important emerg-ing issue? My contention is that these events and actions are signs of a general marketplace movement toward sustainable development—a new form of economic development that balances growth and economics with social needs and environmental quality. It is develop-ment that allows the improvement of quality of life for everyone while living within the carrying capacity[17] of our resources and supporting ecosystems. Good evidence is starting to appear that suggests we are reaching the limits of our resources and beginning to tax the capability of our ecological systems to absorb the products of our activities.

Over the last several years I have served in a unique role at CH2M HILL, looking for new markets and technologies that would be valuable to the firm three to five years hence. In that role, I represented the firm in several global business organizations and professional societies, and I have worked on projects and business development efforts around the world. From this experience, I have gained some valuable insights on the trends and marketplace forces influ-encing the behavior and activities of major corporations and government organizations.

What I have observed is a confluence of trends and market drivers, the organizing theme of which is sustainable development. These are having significant impacts on the structure and operations of public and private sector organizations worldwide. Today there are forces in the market that are driving both public and private sector organizations to rethink how they manage and expand their operations. Stakeholders—employees, customers, clients, public interest groups—are seeing their quality of life deteriorate and are taking those concerns to the marketplace. Not only are these forces reshaping organizational behavior, but they are creating new markets and providing new opportunities for differentiation and innovation.

Recognizing this trend and its effect on current and future business, many companies are making public commitments to the principles of sustainability. They are issuing public reports that reveal not only their financial performance, but their environmental and social performance as well. Environmental management systems are being strengthened and expanded to build consistency and credibility into a company's global operations. Perceived shortfalls in stakeholder expectations are being answered with new environmental and infrastructure projects, built to improve performance.

The Three Worlds

To understand the trends and driving forces that contribute to the movement toward sustainability, it is useful to divide the nations of the world into three broad groups, based on their level of affluence.

World 1. This world consists of the affluent and advanced nations with a collective population of 1 billion people. In these countries economic growth is slow. Its population is aging, consisting mainly of affluent consumers using a disproportionate amount of resources per capita. World 1 nations include the United States, Canada, most of the nations in the European Union, and Japan. All have strong governance frameworks.

World 2. The nations of World 2 are the developing nations characterized by high population growth, with a collective population of 4 billion. Wold 2 nations represent a huge market for goods and services to a young population aspiring to a better quality of life. However, the framework conditions are variable, with wide disparities between the rich and poor. World 2 nations include China, India, Pakistan, Brazil, Venezuela, and many other countries in Central and South America.

World 3. These are the destitute nations, composed of 1 billion people, a young population basically in survival mode. The nations of World 3 lack many of the basic necessities such as sufficient food, shelter, access to freshwater and sanitation. As a result, there is extensive poverty and disease. The framework conditions are poor. The people survive in large part by consuming both renewable and non-renewable resources and harming carrying capacity in the process. World 3 nations include countries in sub-Saharan Africa and Southeast Asia.

Source: Joseph F. Coates, John B. Mahaffie, and Andy Hines, *2025: Scenarios of U.S. and Global Society Reshaped by Science and Technology*, Oakhill Press, (Greensboro, North Carolina. 1997).

Municipal and state governments and federal government agencies are taking a cue from business. Cities throughout the United States are declaring themselves to be "sustainable cities" and are writing that declaration into growth plans and building guidelines. They are doing so, not only because it improves the quality of the city, but because it makes economic sense. Are these cities serious about sustainable development? Many of them are. While some have confined their attention to specific issues, cities like Seattle, Washington; Boulder, Colorado; Scottsdale, Arizona; San Jose, California; and Portland, Oregon, have plans, programs, and policies in place along with sustainability indicators to measure their progress.[18]

The federal government's General Services Administration incorporates sustainability into its building design criteria for federal facilities. The Department of Defense (DOD) is designing its facilities with sustainability criteria and incorporating those criteria into its proposal requests for engineering services. Why is the DOD interested? As the steward of 25 million acres of land and the facilities contained thereon, the DOD is concerned about conserving resources and saving money. In addition, issues like urban sprawl and encroachment can affect its mission.

Other federal agencies, including the National Park Service, the Department of Energy, the Department of State, and the U.S. Environmental Protection Agency, are all advertising for planning and engineering services in sustainable development.[19]

What is sustainable development?

Sustainable development is a goal. It is based on the premise that society's current approach to economic development cannot be sustained much longer at the present rate of consumption. That is, the rate at which we extract and harvest our natural resources and the methods and extent to which we produce, deliver, and consume goods and services are pushing or exceeding the limits of the world's resources and carrying capacity.

This premise is based on sound, emerging evidence showing that the resources and ecological services—the building blocks of our economy—are being used up or damaged to an extent and scale never before imagined and in ways that may not be recoverable. Furthermore, this is happening at all scales, from local pollution of streams and lakes, to poor air quality, to lack of access to freshwater in the underdeveloped nations, to the spread of endocrine disruptors, to global climate change. The consequences of this damage are enormous. If we continue on our current course, that damage will worsen to such a degree that we will be jeopardizing the chances of future generations to enhance or even maintain their quality of life.

How long can we continue on this course? For the United States and other World 1 countries with powerful economies and technological capabilities, resource shortages and ecological damage are generally seen as temporary inconveniences, or they are dismissed as the inevitable consequence of economic development—part of man's conquest of nature. For the developing world, particularly the World 3 countries, it is a different story. Maintaining the health of their resources and ecological services is a matter of survival.

At the beginning of the 20th century, the human population wasn't large enough, nor its technologies strong enough, to have any significant or long-lasting effect on the resources or ecological systems at a world scale. Effects, if any, were localized and mostly short lived. Furthermore, predicted shortages of critical resources turned out to be unfounded or solved technologically by the invention of substitutes. Today, in the first decade of the 21st century, that situation is no longer true. The global population has just passed 6 billion. Furthermore, it is expected to rise to 8 billion over the next 25 years, with essentially all of the growth occurring in the nations of Worlds 2 and 3.

With the concurrent expansion of economic growth, many nations, especially those in the developing world, are about to expand their resource consumption rates by a factor of five or more over the next 30 to 40 years. They are about to do so using conventional technologies and processes—largely non-sustainable. In what is becoming a borderless world, companies can now produce and sell anything, anywhere, anytime.

Seeing the advantage of scale economies and lower labor and operating costs, companies are expanding their facilities into the developing countries. Developing countries welcome this investment and its attendant economic growth, jobs, and overall improvement in their standards of living. The concern that this pathway is not sustainable is much less important than their drive for economic growth. Matters are much worse in the World 3 nations. In these extremely poor countries, the pressures of poverty force people to sacrifice their local environment and resources to survive.

Then there is the matter of homeland security. Society's ability to provide goods and services depends entirely on the efficiency and effectiveness of its production-consumption systems.

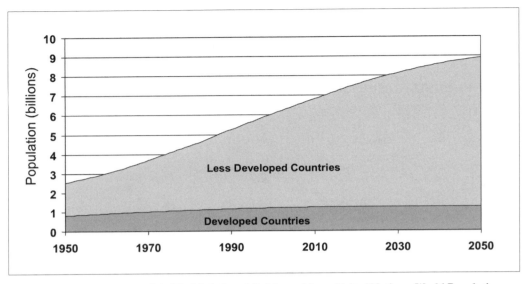

Figure 1-8: Population growth in Worlds 1, 2, and 3. Adapted from United Nations, World Population Prospects: The 2002 Revision, medium scenario. (United Nations, New York, NY, 2003).

The events of September 11, 2001, showed clearly the vulnerabilities—indeed the nonsustainability—of the economies and infrastructure of World 1. Today the entire world faces an entirely new set of vulnerabilities—many, many points of attack against a system designed for efficiency and openness. Whatever solution we devise must include a high degree of protection, redundancy, and robustness.

What is the evidence that we are not sustainable?

In the United States and other World 1 countries, evidence of nonsustainability is beginning to appear:

- *Water shortages.* The growing population in the western United States has been taxing the water supplies of this generally arid climate, so much so that the current five-year drought in the West has forced municipalities to seriously reduce the quantity of water they supply to their residents.

- *Aquifer depletion.* The Ogallala Aquifer, a water source for farmers and residents in the midwestern United States, once held as much water as Lake Huron. Irrigation and other uses have resulted in withdrawal rates that far exceed replenishment rates. Now the aquifer is dry in some regions and continues to be rapidly depleted in others.[20]

- *Dead zone in the Gulf of Mexico.* Agricultural production and the corresponding use of fertilizers and pesticides have created high levels of contaminated runoff. These levels are so high and so widespread that they have created a seasonal dead zone in the Gulf of Mexico. This is a zone of hypoxic conditions, conditions in which oxygen levels are too low to sustain aquatic life. First discovered in the early 1970s, the size and frequency of appearance of this zone have increased substantially. Instead of once every few years, the zone now appears annually, covering an area of over 5,000 square miles.[21] During the past 30 years, the number of oxygen-depleted coastal water bodies has tripled worldwide. Experts relate this trend to increased concentrations of nitrogen and phosphorus in coastal areas, due to wastewater discharges and agricultural runoff. Since 1991, economic losses from these incidents have cost the United States an estimated $280 million.[22]

- *Loss of fishing resources.* Eleven of the 15 most important fishing areas in the world (including the Grand Banks off the coast of Newfoundland) and 70 percent of the major fish species are either fully exploited or overexploited, according to the United Nations (UN) Food and Agricultural Organization.

- *Methylmercury contamination in fish and shellfish.* Levels of contamination have gotten so high that the U.S. Food and Drug Administration and the U.S. Environmental Protection Agency have issued a draft joint advisory warning for pregnant women, nursing mothers. and young children not to eat certain types of fish and shellfish.[24] A total of 68 consumption advisories for mercury contamination have been issued for water bodies in the Gulf of Mexico Program coastal watershed area; 11 of these list estuarine and marine species.[25]

- *Impending oil and gas shortages.* Oil production by some of the largest oil companies has been dropping, and oil and gas reserves are being revised downward. Overall, it now appears that the easily accessible non-Middle East reserves have been exhausted, and the oil companies are having difficulty keeping pace with demand.[26] China, a rapidly industrializing country with a huge population, is adding substantially to world demand.

- *Global climate change.* Emissions of carbon dioxide and other so-called greenhouse gases from energy production, transportation, and other activities are noticeably changing the ambient temperature of the earth. The consequences of global warming include radical climate changes, increased disease, a devastating rise in sea levels, and productivity changes in agriculture. Over the past several decades, insurance losses from large weather-related events have risen from essentially $0 in the 1950s to $9.2 billion a year. The insurance industry is now looking seriously at global climate change as a likely cause.[27] Business is beginning to pay increased attention to the risks and opportunities created by climate change.[28] At this writing, California is drafting regulations for auto manufacturers to cut carbon dioxide emissions by as much as 30 percent in the next 10 years.[29]

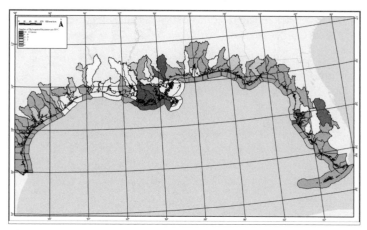

Figure 1-9: Mercury-impaired watersheds in the Gulf of Mexico.
Source: Acke, B.W., J.D Boyle and C.E. Morse, 2000. A Survey of the Occurence of mercury in the Fishery Resources of the Gulf of Mexico. Prepared by Battelle for the U.S. Gulf of Mexico Program, Stennis Space Center, MS. EPA 885-R-00-001 January 2000.

- *Coral reef degradation.* Over one-half of the world's coral reefs are being seriously degraded. Coral reefs offer significant ecological value. They provide a huge coastline habitat for many species of fish as well as protection of coastlines from storms and other sea damage. They also are a major source of income from tourism.[30]

- *Persistent organic pollutants.* There is now evidence of long-range transport of persistent organic pollutants (POPs)—chemical substances that persist in the environment, bioaccumulate through the food chain, and pose a risk of causing adverse effects to human health and the environment. These highly stable compounds can last for years or decades before breaking down. They circulate globally through a process known as the *grasshopper effect.* POPs released in one part of the world can, through a repeated (and often seasonal) process of evaporation, deposition, evaporation, and deposition, be transported through the atmosphere to regions far away from the original source.[31]

While these examples illustrate the sorts of global problems more closely associated with World 1 economies, others serve to illustrate the more severe and near-term threats faced by the World 2 and 3 nations:

- *Access to safe water and sanitation.* Today in the developing world, over 1 billion people do not have access to safe water. Two and a half billion lack access to basic sanitary facilities. By 2025, half the world's population will experience water shortages because of lack of access to safe water.[32]

- *Spread of HIV/AIDS.* HIV/AIDS has reduced the life expectancy in sub-Saharan Africa by 15 years, to age 47. It is pushing households deep into poverty, breaking up families, and leaving millions of orphans.[33]

- *Urbanization and the rise of megacities.* Over the next 25 years, world population will increase by nearly 2 billion people. Ninety-nine percent of that growth will be in the World 2 and 3 nations. Trends in population growth and migration to urban areas are creating megacities in the developing world, cities with huge infrastructure needs and no means to deliver them. Problems include water shortages, lack of sanitation, and exposures to natural disasters.

Figure 1-10: Women villagers in Zambougou, Mali, using a foot pump to obtain fresh water. Photo courtesy of Engineers Without Borders-USA.

- *Deforestation.* In 1998, deforestation triggered flooding in China's Yangtze River basin, killing 3,700 people, dislocating 223 million others, and flooding 60 million acres of agricultural land. The losses are estimated at $30 billion.[34]

These are just a few examples occurring in many places and at many scales. In one way or another, they all reveal resources and infrastructure under stress, and they signal the need for changes in the way we manage them. Sustainable development advocates characterize these incidents as examples of "hitting the wall," that is, incidents with serious consequences that occur when the demands created by the needs and wants of an increasing population exceed the available resources and ecological carrying capacity. The higher the demand, the higher the incident frequency and intensity.

Response by industry and government

Companies like Dow, DuPont, Ford, Nike, General Motors, STMicroelectronics, and many, many others see that their future prospects for growth and profitability are tied directly to their proper stewardship of the environment and their observance of societal values. These companies are making public commitments to sustainable performance and incorporating its precepts into their corporate strategies and values. Furthermore, these companies are devising new products and services to fill newly discovered "white space" opportunities—opportunities to extend their core competencies into new markets resulting from sustainability market drivers.[35] As a direct consequence, new markets are opening up continually to engineering, management consulting, architectural, and accounting firms to help clients become more sustainable in the eyes of their stakeholders. However, with few exceptions, these business opportunities have slipped past most of the engineering community, whose members either have not analyzed the trends sufficiently or have dismissed them as being faddish or even antibusiness.

Business opportunities in sustainable development

Accounting firms like Deloitte Touche Tohmatsu are leveraging their core competencies in financial performance reporting. They are finding these new white space opportunities in helping companies prepare sustainability reports, extending their business beyond traditional financial reporting into environmental and social performance reporting. To these ends, they have created an entirely new practice in sustainable development reporting. They are collecting, analyzing, and developing both internal and external reports on a vast array of company performance information. Through these activities they gain tremendous insights into a company's performance shortfalls, thus positioning themselves to manage or perhaps take on the follow-on engineering work.

Nike, the global sports and fitness company, has made a strong commitment to sustainable development, extending to all parts of the organization and its suppliers. To these ends, it has engaged several organizations including Business for Social Responsibility and CH2M HILL to establish global voluntary water-quality standards for its suppliers. In addition, the company is switching over to natural fibers for its apparel products, making it one of the world's largest customers for organically grown cotton. It also established a recycling program for athletic shoes. Called Reuse-A-Shoe, the company collects manufacturing rejects and post-consumer discards, processing the material into athletic surfaces. Its stated aim is to raise the recycling level so that old shoes can be used to make new ones.

Nike is also moving toward zero use of toxic materials, phasing out polyvinyl chloride from its products and eliminating volatile organic compounds from its manufacturing processes. Since 1995, Nike has reduced solvent use by 89 percent in its Asian manufacturing plants, moving to water-based formulas.

Nike's commitment to change was born out of an incident in the mid-1990s. In 1996, *Life* magazine published an article declaring that one of the contractors to Nike was using child labor in Asia in the manufacture of soccer balls. In the flurry of criticism that followed, Nike quickly recognized that such practices were not only poor from an ethical standpoint, but devastating to a company whose image and reputation are key factors to its sales. Since then, Nike has made a commitment to sustainability, examining closely the impact of its products and operations on the environment, and seeking to improve the lives and working conditions of its workers. The company produces a community investment report detailing its investments in the communities in which it operates. It also produces a corporate responsibility report that assesses the impacts caused by its own operations. The report includes additional reports on the company's efforts to achieve sustainability and to understand and manage global labor compliance, as well as how the company is involved in local communities.[36]

Designing sustainable products

In addition to his work at Ford's River Rouge plant, architect William McDonough, along with his partner, chemist Michael Braungart, are changing the way in which products and facilities are designed. Using the design principle they have termed *Cradle to Cradle*, their company, McDonough Braungart Design Chemistry, LLC (MBDC), works with clients to produce new products that are not only environmentally safe, but totally recyclable. Cradle to Cradle Design extends by-product synergy to its logical endpoint. In this concept, there is no waste. All wastes become either feedstocks to another process or are returned to the environment as nutrients. Even their new book, *Cradle to Cradle*, is made of a polymer material that can be completely recycled, that is, broken down and remade into another book or other products indefinitely.[37] Another example of their Cradle to Cradle Design success is the upholstery fabric Climatex Lifecycle. This fabric is a combination of wool and ramie (a plant material), which is safe for both the public and the environment. The process waste is even used by garden clubs for mulch.[38]

In *Cradle to Cradle*, McDonough and Braungart advance the concept of eco-effectiveness, in which design emulates nature. They conceive of two systems—metabolisms as they call them—each having a distinct character and function. In the biological system, materials are selected based on their contribution to the biological metabolism, that is, based on their eco-logical safety and ability to biodegrade. In the other system, the technical system, materials valuable for their performance qualities, but that may be toxic and normally are nonrenew-able, are kept within the technical system and are continually recovered and reused. The concept of eco-effectiveness is discussed in more detail in chapter 2.

Turning to civil and environmental engineering, McDonough poses an interesting challenge.

We are proposing a new design assignment where people and industries set out to create the following:

- *Buildings that, like trees, are net energy exporters, produce more energy than they consume, accrue and store solar energy, and purify their own waste water and release it slowly in a purer form.*

- *Factory effluent water that is cleaner than the influent.*

- *Products that, when their useful life is over, do not become useless waste, but can be tossed onto the ground to decompose and become food for plants and animals, rebuilding soil; or, alternately, return to industrial cycles to supply high quality raw materials for new products.*

- *Billions, even trillions of dollars worth of materials accrued for human and natural purposes each year.*

- *A world of abundance, not one of limits, pollution, and waste.[39]*

Engineers, clearly accomplished in matters of design, may ask, If that's the new design as-signment, what was the old design assignment? Is there a problem? McDonough replies in this way.

Consider looking at the industrial revolution of the 19th century and its aftermath as a kind of retroactive design assignment, focusing on some of its unintended, questionable effects. The assignment might sound like this. Design a system of production that:

- *Puts billions of pounds of toxic material into the air, water, and soil every year*

- *Produces some materials so dangerous they will require constant vigilance by future generations*

- *Results in gigantic amounts of waste*

- *Puts valuable materials in holes all over the planet, where they can never be retrieved*

- *Requires thousands of complex regulations to keep people and natural systems from being poisoned too quickly*

- *Measures productivity by how few people are working*

- *Creates prosperity by digging up or cutting down natural resources and then burying or burning them*

- *Erodes the diversity of species and cultural practices[40]*

Large pipes, small motors, big savings

Amory B. Lovins brings the engineering design paradigm problem closer to home. Lovins is the CEO of the Rocky Mountain Institute (RMI) and co-author of the book *Natural Capitalism*. His primary focus is on energy and resource efficiency in a broad cross section of industry sectors ranging from automobile, real estate, electricity, water, semiconductor, to several others. The *Wall Street Journal* named Lovins one of 39 people worldwide "most likely to change the course of business in the '90s"; *Newsweek* has praised him as "one of the Western world's most influential energy thinkers"; and *Car* magazine ranked him the 22nd most powerful person in the global automotive industry.[41]

Lovins is very effective in introducing radical ideas and insights into engineering, many so simple and obvious that it makes one wonder how they ever could have been overlooked. For example, in *Natural Capitalism*, Lovins and his co-authors use an example of industrial piping design, noting that in traditional engineering pipe sizing is determined by comparing the cost of larger (fatter) pipe that uses less energy with the value of the pumping energy saved. The traditional approach ignores the capital costs of the pumps and the related equipment. Using fatter pipes allows the design engineer to specify smaller pumps and motors (friction decreases by about the fifth power of the pipe diameter), saving considerable capital and operating costs. More savings in money and energy can be gained by designing the pipe layout before placing the industrial equipment. Shorter, straighter pipe runs are more energy efficient and easier to maintain.[42]

Some years ago, I had the privilege of meeting Lovins as he was passing through Denver on the way to Snowmass, Colorado, the home of his organization. In the course of the conversation, Lovins related a story about a meeting he and his people had with Admiral David Nash, then head of the Naval Facilities Engineering Command (NAVFAC). It seems that Admiral Nash was concerned with the cost of constructing and operating the navy's facilities. He asked Lovins to review the design of the facilities and see what could be done to reduce costs. Lovins and his team came back with a long list of changes, most of which could be traced back to the engineering design. In many cases, the designers had failed to take the cost of energy into account. Their design paradigm was to have the position of the large equipment drive the design, with the location and routing of fluids piping being treated as an afterthought. As a result, the facility piping was burdened with small-diameter pipes connected with numerous elbow joints, all of which increase friction and use more energy.

As Lovins tells it, he and his team were presenting page after page of instances of energy-wasting designs in the navy's facilities when the admiral stopped them in midsentence.

The admiral turned to his executive officer and asked, "Do we have a list of the engineering firms that designed these facilities?" The officer said that yes, he did have such a list and proceeded to pull it from his file.

"Good," the admiral said. "They'll never work for us again!"

The admiral then turned back to Lovins and asked if his team had the capability to design navy facilities incorporating his group's energy- and money-saving techniques. Lovins, of course, answered yes.[43]

Under Admiral Nash, the Naval Facilities Engineering Command changed its protocols for its facility design procurements. Instead of selecting design firms based in part on lowest first cost, the new rules incorporated life cycle cost as both a criterion and an incentive. In this innovative approach, the navy would select firms based in part on lowest life cycle cost. As an incentive, the navy would share with the designer the cost savings achieved over and above conventionally designed facilities.

Amory Lovins and his team also did a similar review for Interface Corporation, whose CEO, Ray Anderson, has been a leader and a major force behind corporate sustainable development. Again the team turned up similar design problems, the same kinds of problems found at NAVFAC facilities. Sometime later, I spoke to Jim Hartzfeld, a senior manager at Interface, who worked with Lovins on the review. Hartzfeld said that he came into the review thinking that Lovins and his team would simply acknowledge Interface's good work, since the company was a known leader in sustainability. However, Lovins's team uncovered many design and energy inefficiencies. Hartzfeld's reaction was expected: "When I saw what the RMI team came up with I felt like I was hit with a 2 by 4!"[44]

What is the business of business?

Sustainability is moving both business and government agencies outside the traditional boundaries of organizational responsibility. Economist Milton Friedman's old axiom, "The business of business is business," is no longer true. Corporate responsibilities are expanding outside the traditional financial realm and into areas of environmental and social responsibility, often beyond the laws and regulatory requirements. Furthermore, these responsibilities are extending up and down the value chain. Companies and government agencies, in response to stakeholder pressures, are imposing sustainabil-

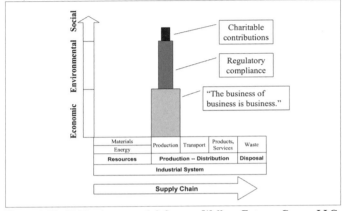

Figure 1-11: Old business model. Source: Wallace Futures Group, LLC.

ity requirements on their suppliers to provide raw materials harvested or produced under sustainability codes. They are also developing recyclable products and packaging or starting to take back products that are beyond their useful life. Product take-back laws are beginning to take hold in Europe. In the United States, computers and other electronic equipment have become the focus of take-back campaigns because of the large volume of so-called e-waste being generated. Twenty states are now considering some form of take-back legislation.[45] In May 2004, Dell and Hewlett-Packard announced their support for computer recycling and promised to assume more of the cost of recycling used computers.[46]

From a strict business point of view, the realization that our current form of economic development is not sustainable is starting to affect how industry and government operate. People are beginning to understand and trace apparent degradations in their quality of life back to nonsustainable behavior. More important, they are making their concerns known in both the mar-

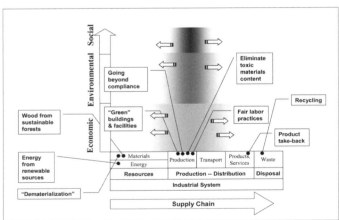

Figure 1-12: New business model. Source: Wallace Futures Group, LLC.

16

ketplace and the voting booth. In response, industry and government leaders are making changes to their plans and operations. Many are treating these relatively new concerns not as threats, but as opportunities for differentiation and as a catalyst for innovation. Still, not all companies and government agencies are embracing sustainable development. Consideration of sustainability principles appears, at this point, to be a function of the impact of their operations on the environment combined with stakeholder pressures.

The journey toward sustainability: Where do we stand today?

Although there is a compelling case for moving our extraction, production, and consumption systems and infrastructure toward sustainable development, we still have a long way to go. Not all organizations, whether public or private, have been directly and substantially affected by the problems associated with nonsustainable behavior. Those that have may not have read much into the experience. To others, sustainable development appears as just another twist on the perennial environmental theme.

In summary, here is the situation today regarding the case for sustainable development:

1. The evidence of nonsustainable conditions is strong, but it is not sufficiently comprehensive or defendable, nor does it create a high sense of urgency.

It is clear that resource and ecological problems exist in may parts of the world, including the United States. As noted earlier in this chapter, respected organizations have calculated and verified that at the world scale, 1 billion people do not have access to clean water, and 2.5 billion people do not have access to sanitation. In addition, fishing resources have been severely depleted, and water resources are drying up due to overuse.

Although the list of problems goes on, it is still not complete or comprehensive enough to argue the larger case that our overall model for economic growth is not sustainable. Much of the information, while certainly alarming, is still anecdotal. Furthermore, it is not part of a systematically collected and verified package that points directly to an overall problem of nonsustainability. Critics can still muster a strong argument that these predictions of doom have been heard before, and that history tells us that they have turned out to be false. Projections of resource shortages are frequently rescinded soon after by new resource discoveries or by invention of replacements. Ecological disasters have been avoided or repaired with little long-term effects.

Further contributing to this doubt is the issue of global climate change. Projected change in the world's climate caused by the emission of greenhouse gases is rated by thousands of world-renowned scientists as an incredibly serious problem, worthy of immediate worldwide action. Yet there is a small but well-respected group of experts who say that there is no proof that such a condition exists. As a result, relatively little action has been taken, and the clash between scientific experts over this matter has raised doubt in the minds of the stakeholders: If the supposed experts cannot agree on this important issue, then what other of the so-called problems may be suspect?

Criticism over the judgment that we are in the midst of an ecological crisis is not confined to global climate change. Professor Björn Lomborg, in his book *The Skeptical Environmentalist*, argues that the statistics used in many of the popular and respected assessments are flawed, giving rise to the argument that things are not as bad as they are made out to be. Lomborg has delved deep into the assessments and turns up a number of supposedly flawed assumptions and contradictions. When looked at carefully, many of the doomsday predictions, for example, the loss of forests, turn out to be just the opposite.[47] In addition, the book *Earth Report 2000*[48] takes issue with many of the statements and predictions of the Worldwatch Institute, a respected nonprofit organization that researches environmental and resource trends.

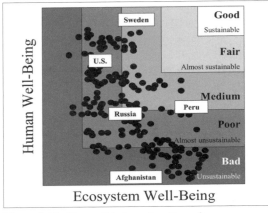

Figure 1-13: **The well-being of nations.** Source: Adapted from robert Prescott-Allen, *The Wellbeing of Nations* (Washington, DC, Island Press. 2001).

Other critics, cloaked in righteous skepticism, are using the doubts raised as sufficient reason for maintaining the status quo. Sustainable development, they argue, is part of a plot to enhance the economy of other countries at the expense of ours. Still others start from the unsupportable premise that we live in a kind of "cornucopia world," one in which the resources and carrying capacity are, for all intents and purposes, boundless. In this world, shortages are not real, but simply the result of poor management in a governance structure that is not democratized or market based. Some of that may be true; however, it is very difficult to convince those who are suffering from the consequences of the shortages. It would be instructional to change the premise: What if we were not living in a cornucopia world but rather in a "zero sum game world," in which the success of a few nations is being achieved at the expense of others? World 1 nations are in fact using many times more resources to achieve their level of development and quality of life than those used by the developing nations.

The net result is that we have a growing collection of anecdotes, individually strong but collectively insufficient to challenge the status quo. Doubts about the veracity of the individual elements, reinforced by the self-serving agendas of others, have created an interesting problem set, but one that lacks a high sense of urgency to address.

2. Work has been started at the global level to systematically assess the problem.

In the face of doubt and controversy, a number of organizations have been systematically gathering information in order to make a comprehensive assessment of the current state of the world. The most promising in terms of its coverage and consensus agreement appears to be the Millennium Ecosystem Assessment. Results are expected in early 2005.

The key reason for conducting this level of assessment is to enable leaders in both government and industry to make informed decisions about which problems to address and how to address them. Today, without a comprehensive and defendable assessment, decisions about how to best move toward sustainable development are left to the whims and agendas of the groups that wield the most public influence.

Other assessment efforts include the annual *State of the World* publication of the Worldwatch Institute. Worldwatch publications tend to view the situation as negative. However, over the years it has assembled an extensive database of ecological indicators that are useful in making one's own assessments.

An important contribution to global environmental and social assessment is Robert Prescott-Allen's report *The Wellbeing of Nations*.[49] Using a wide range of data sources, the author developed a country-by-country index of quality of life and the environment—the Wellbeing Assessment—and displayed the results qualitatively in terms of human and ecosystem well-being. In the author's judgment, no country has achieved sustainability, nor can any even be classified as "good," in terms of both human and ecosystem well-being. Furthermore, the author points out that to improve human well-being, countries will likely do so at the expense of ecosystem health.

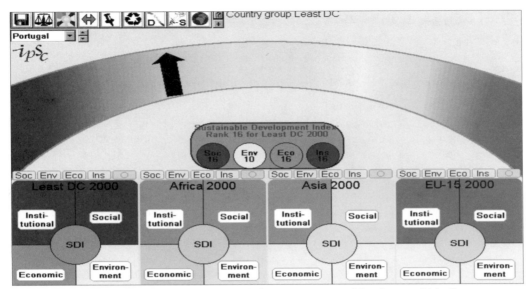

Figure 1-14: The "Dashboard of Sustainability" is free, non-commercial software which allows to present complex relationships between economic, social and environmental issues in a highly communicative format aimed at decision-makers and citizens interested in Sustainable Development. Illustration courtesy of IIDS's Consultative Group On Sustainable Development Indicators (CGSDI).

An interesting piece of software is the Dashboard Tool, developed by the Consultative Group on Sustainable Development Indices. The tool takes approximately 100 social, economic, and environmental indicators from a number of international sources and combines them to make a qualitative, country-by-country assessment of sustainability. All of this information is displayed on an attractive software interface using a dashboard metaphor. The International Institute for Sustainable Development in Winnipeg, Canada, is currently supporting this activity. This tool is available on the Internet through the CD-ROM that accompanies this book.

3. Even in the face of doubt, industry and governments are beginning to take action, primarily due to stakeholder pressures and issues of competitiveness and innovation.

In spite of the lack of definitive and decisive information about the trends and consequences of society's overall nonsustainable behavior, industry and governments are beginning to take action on specific issues. At the local level in the United States, issues such as urban sprawl, water shortages, open space, recycling, air and water pollution, hazardous waste, jobs, and unfair labor practices are being addressed to various degrees. In a growing number of cases, these issues are starting to be addressed collectively under the rubric of sustainable cities programs. Here, government officials and community leaders have recognized these issues as symptoms of a growing citizen concern over quality of life. At the regional and global scales, organizations and national governments are setting goals and priorities on the problems and issues identified in the UN's Agenda 21 plan of action.

Industry is responding too, primarily because of stakeholder pressures and fear of loss of reputation. However, some companies see the concerns over sustainable development as opportunities for differentiation in the marketplace: ways to meet previously unrecognized customer attitudes and needs. Other companies consider the issue of sustainable development as a catalyst for innovation, not only by opening up new markets, but as a new approach for creating products and services.

What needs to be done

What is the best approach for making progress toward sustainable development? Based on the current situation described above, I see four steps that must be taken:

1. Develop a credible, defensible assessment of the problem at multiple scales.

At this writing, there is no common agreement as to the scope and extent of the problems—current and potential—associated with nonsustainable development. The absence of such an agreement creates a void, giving rise to a multiplicity of views, often conflicting, on which problems are the most important.

Furthermore, there is no agreement as to the principles that should be followed in order to address those problems. The primary issue here is what level of proof is needed before action is taken. Agenda 21, the call to action published in the Rio Declaration, calls for the application of the *Precautionary Principle* in making these determinations. The Rio Declaration states:

> *In order to protect the environment, the precautionary approach shall be widely applied by States according to their capabilities. Where there are threats of serious or irreversible damage, lack of full scientific certainty shall not be used as a reason for postponing cost-effective measures to prevent environmental degradation.* [30]

The possibility of depletion and irreversible damage to the world's critical resources and ecosystems suggests that the Precautionary Principle ought to be applied to future actions.

To date, the most promising effort currently under way that will provide this information is the previously mentioned Millennium Ecosystem Assessment. As presented by the organizers, the assessment appears to be broad and thorough enough to begin to fill the existing information voids as well as able to identify key information gaps that need to be addressed with follow-on work.

Hopefully, the assessment will cut through the vast array of conflicting information and opinions about the state of the world and how best to achieve sustainability.

2. Devise agreed-upon measures of progress toward sustainability.

If the assessment is successful in meeting its objectives, it will provide sound, comprehensive information about the current and potential future condition of the earth's ecosystems, across multiple scales. This information will form the basis for measuring progress toward sustainable development and determining which problems deserve priority attention.

Armed with this information, stakeholders will be able to reduce the vast array of sustainable development measures down to a manageable set, ordered by priority. In the face of scarce resources, it is obviously important to address the most critical problems first.

3. Create an *environment for innovation.*

Moving toward sustainable development will be an exercise never before attempted in the history of the world. It will be a messy, nonlinear process, filled with fits and starts. As stated earlier, it will require a more or less total overhaul of the current production-consumption system, together with an unprecedented investment in infrastructure and in new technologies, most of which have yet to be invented.

To accomplish this overhaul, we must create an *environment for innovation*, an environment in which the participants understand both the urgency of the problems of nonsustainability and the requirements for unleashing the creative talents and instruments to solve them. The characteristics of this environment include (1) easy sharing of information across organizational and national boundaries, (2) an atmosphere of trust, and (3) diversity of participation from multiple cultures and disciplines.

Engineers' role in sustainable development

While it is clear that engineering firms will have a role in sustainable development, it is not yet clear what that role will be. In-house engineering departments, "icon" consultants (such as Paul Hawken, Amory Lovins, L. Hunter Lovins, Bill McDonough, Brian Nattrass, and Mary Altomare), management consultants, accounting firms, and architectural firms that design green buildings are currently providing most sustainable engineering services. Accounting and management consulting firms are entering this market by leveraging their environmental management services and offering strategic advice on greening the company. Although architects are incorporating sustainability into their designs, they typically use "natural" materials and off-the-shelf technologies with green attributes, rather than address the systems aspects of sustainability. Accounting firms are engaged in the preparation of environmental or sustainable development reports that collect, document, and present information on the economic, environmental, and social aspects of a company's performance.

Today engineering firms generally are not invited to participate in these projects unless the projects have been broken out into pieces, handing to the engineering firm those elements that require the application of the more conventional technologies. There are a number of reasons for this. This sector is viewed as steeped in traditional practices, lacking the necessary tools to integrate systems across disciplines so that they will work properly. Engineering firms have a low comfort level with trying out new materials and systems, primarily because of an innate fear of litigation and an inability to sell innovation to clients. Engineering firms also tend to organize around specific market sectors, for example, transportation, mining, forest products, or chemicals. Each sector tends to view the world through a single market-sector lens and does not have the time or the inclination to learn from the experiences of other groups.

Some engineering firms have created sustainable development practices, but most have yet to define sustainability, establish performance standards, or assemble the requisite skills to meet client needs and deliver true progress toward sustainability. All too often projects are loaded up with environmental features and passed off as "sustainable" without analyzing whether those features will, from a systems view, make any true contribution to sustainability. Additionally, some firms see the subsets of the sustainability movement (e.g., smart-growth initiatives) as threats to their business and have actively sought to defeat such initiatives.

As a prescription to this problem I see four parts:

1. Become part of the solution, not part of the problem.

For engineers, McDonough's look at what he calls the "retrospective design assignment" ought to be chilling.[31] Some of the points he raises about the poor design of our system of production are the things that, for many decades, engineers have taken pride in designing and building. Pollution control systems that meet, or even exceed, regulations still release substantial quantities of pollutants into the air, water, and soil and tax ecological carrying capacity. Increasing the cost efficiency of harvesting or extraction systems makes it cheaper to access increasingly scarce resources. Designing a system that lowers labor costs still forces workers into the job market with skills incompatible with current labor needs. Furthermore, these sorts of projects have been done so many times that the work has become a commodity and priced accordingly.

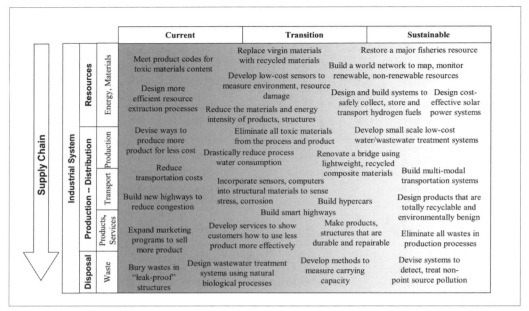

Supply Chain / Industrial System	Current	Transition	Sustainable
Resources — Energy, Materials	Meet product codes for toxic materials content; Design more efficient resource extraction processes	Replace virgin materials with recycled materials; Develop low-cost sensors to measure environment, resource damage; Reduce the materials and energy intensity of products, structures	Restore a major fisheries resource; Build a world network to map, monitor renewable, non-renewable resources; Design and build systems to safely collect, store and transport hydrogen fuels; Design cost-effective solar power systems
Production – Distribution — Production	Devise ways to produce more product for less cost	Eliminate all toxic materials from the process and product; Drastically reduce process water consumption	Develop small scale low-cost water/wastewater treatment systems
Production – Distribution — Transport	Reduce transportation costs; Build new highways to reduce congestion	Renovate a bridge using lightweight, recycled composite materials; Incorporate sensors, computers into structural materials to sense stress, corrosion; Build hypercars; Build smart highways	Build multi-modal transportation systems; Design products that are totally recyclable and environmentally benign
Production – Distribution — Products, Services	Expand marketing programs to sell more product	Develop services to show customers how to use less product more effectively; Make products, structures that are durable and repairable	Eliminate all wastes in production processes
Disposal — Waste	Bury wastes in "leak-proof" structures; Design wastewater treatment systems using natural biological processes	Develop methods to measure carrying capacity	Devise systems to detect, treat non-point source pollution

Figure 1-15: Engineers' new design assignments. Source: Wallace Futures Group, LLC.

If Bill McDonough's retrospective does not scare you, then Amory Lovins's whole-system design approach should. Imagine being able to go at will to almost any building and industrial facility as Lovins and his teams do and uncover dozens of design flaws, all of which cost the owner-operator substantial amounts of money, and all of which could have been cured by a simple change in design mindset by the engineer.

Becoming part of the solution will create entirely new challenges and opportunities for engineers and scientists to apply their knowledge and creativity to a new set of problems, or perhaps to old problems recast in a new way. For example, instead of working on ways to harvest or extract more natural resources at a lower cost, we can be looking for ways to increase the productivity of natural resources, that is, doing more with less. We can also find ways to extract valuable materials from landfills or other disposal sites. Instead of figuring out how to sell more product, we can be selling higher-margin services, helping the customer use less product more effectively to satisfy the customer's needs. Livio DeSimone and Frank Popoff framed these thoughts as *eco-efficiency*, essentially doing more with less and creating less waste in the process.[32] Eco-efficiency is discussed in detail in chapter 2.

The transition to sustainability will be marked by an array of engineering projects designed to support or implement eco-efficiency principles. Examples of these projects are depicted in Figure 1-15. These projects offer new and exciting challenges and require new knowledge and skills. Accordingly, they can claim higher margins. Furthermore, these are the kinds of projects that excite young, creative engineers to join your company.

2. Develop and communicate the business case for sustainable development to clients.

For business as well as government, making a commitment to sustainable development starts at the highest level of the organization. Here, business and government leaders, faced with increasing competition, cost pressures, and a rapidly changing environment, need convincing arguments that becoming more sustainable will contribute to accomplishing the mission and vision of their respective organizations. Furthermore, to be successful, that contribution must be converted into effective strategies and tactics and approved and implemented throughout the organization.

To date, clients have relied on management consulting firms or icon consultants to create a vision for the organization regarding sustainable development. Unless an engineering firm can claim this level of expertise, it is unlikely that it will be called upon to help a company in this way. But what engineering firms can offer these clients is a unique perspective on sustainable development.

Typically, engineering firms have worked for clients in many different industrial and government sectors. Each of these sectors is uniquely affected by the trends and market drivers pertaining to sustainable development. Thus an engineering firm with multisector experience can frame the issue of sustainable development in terms of how these trends have affected other sectors, including those of the client. Also, engineering firms can offer sound, practical advice on how sustainable development practices and technologies could be implemented successfully at a client's facility.

For engineering firms, a business case for sustainable development can be made at the project level, by showing successful applications of sustainability principles to projects. Having a portfolio of successful sustainable development projects will strengthen the firm's competitive position for future work.

3. Visualize and deliver projects that truly improve sustainable development conditions.

The transition toward sustainable development will be accomplished through a systematic infrastructure overhaul, replacing regularly nonsustainable processes, systems, and technologies with those that better meet sustainability criteria. In the best of circumstances, each project should make a net contribution toward the sustainability goals. This contribution would be measured in terms of such things as reductions in energy and critical material usage, less or zero toxic release, higher durability, higher service intensity, and lower life cycle cost. To help ensure progress, performance of the design should be evaluated against a set of sustainability goals and indicators. These goals and their respective indicators need to be comprehensive enough to ensure that net sustainability performance has been maintained or improved. In other words, the project should make an overall contribution to sustainability. Performance gains in one dimension of sustainability should not be delivered at the expense of others.

Moving toward sustainability will require not only the development and application of new individual technologies to projects, but the integration of those technologies to one another in a way that allows them to function cohesively. While a technology may be able to achieve a high level of performance on its own, that may not be the case when it is integrated with other technologies and systems. Here, benchmarking against what others have achieved will be an important tool. Practitioners may learn through the success of others, but they will be encouraged to set new goals for levels of performance. Procedures for setting project sustainability goals and indicators to measure performance are discussed in detail in chapter 6.

4. Test and verify technical performance.

As appliers of sustainable technology, engineers ought to organize and operate programs to systematically test and verify technology performance. Furthermore, this performance needs to be defined in term of sustainable development criteria. Since sustainable development is an emerging field of endeavor, the criteria are changing frequently.

Sustainable development technologies have three important characteristics that make systematic testing and verification extremely important.

- *Technology effectiveness.* Many of the technologies may be effective in some low-key applications but might not stand up under industrial operating conditions.

- *Complexity.* Many of the technologies achieve sustainable performance by employing several techniques, many of which are nonconventional. For example, wastewater treatment using engineered wetlands may be much more complicated to maintain and operate than a conventional wastewater treatment system.

- *Operational differences.* Being nonconventional, the resulting systems need to be well understood so that their operation and maintenance can be transferred to a standard operator, the one who operated the conventional system. Often this operator is unaware of the details of the operation of the new system. Consequently, it is important to carefully investigate all the critical operating parameters and translate those parameters into detailed operating specifications.

As Richard Weingardt points out, "the world is run by those who show up."[33] Right now, the U.S. engineering community is uniquely positioned to take action for the good of the nation as well as for the rest of the world. Not only do we possess the predictive tools to see this impending problem, but we possess the technological tools and creativity to solve it. It seems we have two choices. The first choice is to do what we have always done: tinker with the current production-consumption model hoping to make incremental reactive changes while navigating from crisis to crisis. The second and far better choice is to use our vision and tools to proactively lead the country out of this impending crisis and into a new industrial age of sustainable development.

---

[1] Tom Gladin of the University of Michigan School of Business Administration is quoted on the Dow Corporation Web page "Dow Polyurethane Carpet Backings: Sustainable Development," http://www.dow.com/carpet/susdev/.

[2] "Light Rail's Besieged with Riders! This Is a Problem?" special report by Light Rail Progress, Austin, TX, March 2001, http://www.lightrailnow.org/news/n_den002.htm.

[3] LandVote, www.landvote.org. This site contains a complete list of local and state conservation-related balloting.

[4] LEED refers to the Leadership in Energy and Environmental Design (LEED) Green Building Rating System developed by the U.S. Green Building Council (USGBC). The LEED system is described in more depth in chapter 6.

[5] Charles O. Holliday Jr., Stephan Schmidheiny, and Philip Watts, Walking the Talk, 212.

[6] David Perkins and Michael Wortley, "CemStar Increases Production and Profits," Global Cement and Lime Magazine, June 2003.

[7] Private conversations with Dr. Gordon Forward, Andrew Mangan.

[8] Ibid.

[9] Charles O. Holliday Jr., "Sustainable Growth Progress Report" DuPont, http://www1.dupont.com/NASApp/dupontglobal/corp/index.jsp?page=/content/US/en_US/social/SHE/usa/us1.html.

[10] "DuPont Science Leader Discusses "Sustainability and Integrated Science for the 21st Century" Speech delivered by DuPont Central Research and Development Vice President Dr. Uma Chowdhry at the 2003 AAAS Meeting DENVER, Colo., February 17, 2003. DuPont Press Release.

[11] Scientific Certification Systems, "Specification SCS-EPP11-03a," in Standard Specification for the Evaluation and Certification of Environmentally Preferable Carpet, pilot (Emeryville, CA: Scientific Certification Systems, May 20, 2003), p.4.

[12] Roger Cowe, "No Chip Off the Old Block," Tomorrow, March-April 2001, 30–31.

[13] Kent E. Portney, Taking Sustainable Cities Seriously: Economic Development, the Environment, and Quality of Life in American Cities (Cambridge, MA: The MIT Press, 2003). 70–71.

[14] Scott Johnson, "The Economic Case for High Performance Buildings," Corporate Environmental Strategy 7, no. 4 (2000): 350–361.

[15] Scott Johnson, in discussion with the author, February 2004.

[16] Bill Franzen, in discussion with the author, July 2004. The current construction cost for the newest Poudre School District school is $99 per square foot. Comparable construction costs for a conventionally designed and constructed school building in this area are $120 per square foot.

[17] The carrying capacity of a particular area is defined as the maximum number of a species that can be supported indefinitely by a particular habitat, allowing for seasonal and random changes, without degradation of the environment and without diminishing carrying capacity in the future. Source: Garrett Hardin, 1974. "The rational foundation of conservation." North American Review, 259 (4) :14-17.

[18] Kent E. Portney, Taking Sustainable Cities Seriously, 69.

[19] Roger D. Blevins, "Air Force Sustainable Planning and Development" (presentation at the American Planning Association National Planning Conference, New Orleans, LA, March 14, 2001).

[20] Lester W. Milbrath, Envisioning a Sustainable Society (Albany: State University of New York Press, 1989): 29–30.

[21] Elizabeth Carlisle, "The Gulf of Mexico Dead Zone and Red Tides," http://www.tulane.edu/~bfleury/envirobio/enviroweb/DeadZone.htm.

[22] Lester R. Brown, Michael Renner, and Brian Halweil, Vital Signs 1999 (New York: The Worldwatch Institute / W. W. Norton & Company, 1999), 126.

[23] Lester R. Brown, Christopher Flavin, and Hilary French, The State of the World, 1999 (New York: The Worldwatch Institute / W. W. Norton & Company, 1999), 13.

[24] U.S. Food and Drug Administration and U.S. Environmental Protection Agency, What You Need to Know About Mercury in Fish and ShellFish, EPA-823-R-04-005 (Washington, DC: March 2004).

[25] B. W. Ache, J. D. Boyle, and C. E. Morse, A Survey of the Occurrence of Mercury in the Fishery Resources of the Gulf of Mexico, prepared by Battelle for the U.S. EPA Gulf of Mexico Program (Stennis Space Center, MS: January 2000), xii, http://mo.cr.usgs.gov/gmp/Downloads/newHgFinalReport.zip.

[26] Alex Berenson, "An Oil Enigma: Production Falls Even as Reserves Rise," New York Times, June 12, 2004.

[27] Evan Mills, Eugene Lecomte, and Andrew Peara, U.S. Insurance Industry Perspectives on Global Climate Change, MS 90-4000 (Berkley, CA: Lawrence Berkeley National Laboratory, 2001), 13–14.

[28] Barnaby J. Feder, "Survey Finds More Corporate Attention to Climate Change," New York Times, May 9, 2004.

[29] Danny Hakim, "California Weighs Tighter Fuel Economy," New York Times, June 9, 2004.

[30] World Resource Institute, Sustainable Development Information Service, "Global Trends, Diminishing Returns: Coral Reefs: Assessing the Threat," http://www.wri.org/wri/trends/coral.html.

[31] United Nations Environment Programme, "Governments Finalize Persistent Organic Pollutants Treaty" (press release, Johannesburg, South Africa, December 10, 2000).

[32] Ibid., 11.

[33] World Resource Institute, "Global Trends, Global Opportunity: Trends in Sustainable Development" (published for the United Nations Department of Economic and Social Affairs for the World Summit on Sustainable Development, 2002), 5.

[34] Amory B. Lovins, Hunter L. Lovins, and Paul Hawken, "A Road Map for Natural Capitalism," Harvard Business Review (May–June 1999): 146.

[35] The term white space opportunities was coined by Gary Hamel and C. K. Prahalad in their book Competing for the Future (Boston: Harvard Business School Press, 1994), 229–230.

[36] Nike's reports are available on the nikebiz.com Web site at http://www.nike.com/nikebiz/nikebiz.jhtml?page=29.

[37] William McDonough and Michael Braungart, Cradle to Cradle (New York: North Point Press, 2002), 5.

[38] William McDonough and Michael Braungart, "Cradle to Cradle Design Guidelines," a report for MBDC, 2003, p. 3, http://www.mbdc.com/challenge/Cradle-To-Cradle_Design_Guidelines.pdf .

[39] MBDC, "The Next Industrial Revolution," http://www.mbdc.com/c2c_nir.htm.

[40] Ibid.

[41] Source: Rocky Mountain Institute, Staff bios, http://www.rmi.org/images/other/StaffBios/BioALovins.pdf

[42] Paul Hawkin, Amory Lovins, and L. Hunter Lovins, Natural Capitalism (Boston: Little, Brown and Company, 1999),115–119.

[43] Amory Lovins, in discussion with the author, 1998.

[44] James "Jim" Hartzfeld, in discussion with the author, 1998.

[45] Silicon Valley Toxics Coalition, clean computer campaign, http://www.svtc.org/cleancc/usinit/initsmap.htm.

[46] Laurie J. Flynn, "2 PC Makers Favor Bigger Recycling Roles," New York Times, May 19, 2004.

[47] Björn Lomborg, The Skeptical Environmentalist: Measuring the Real State of the World (New York: Cambridge University Press, 2001).

[48] Ronald Bailey, ed., Earth Report 2000: Revisiting the True State of the Planet (New York: McGraw Hill, 2000).

[49] Robert Prescott-Allen, The Wellbeing of Nations (Washington, DC: Island Press, 2001).

[50] United Nations Department of Economic and Social Affairs, "Rio Declaration on Environment and Development, Principle 15" (presented at the United Nations Conference on Environment and Development, Rio de Janeiro, June 3–14, 1992.)

[51] William McDonough, Michael Braungart, "The Next Industrial Revolution," Atlantic Monthly, Oct 1998, pp. 82-92.

[52] Livio DeSimone and Frank Popoff, Eco-efficiency: The Business Link to Sustainable Development (Cambridge, MA: The MIT Press, 1997).

[53] Richard Weingardt, Forks in the Road (Denver, CO: Palamar Publishing, 1998), 75.

CHAPTER 2

Sustainable Development: Origins, Concepts, and Principles

The concept of sustainable global development is based on two central questions. First, is it possible to increase the basic standard of living of the world's population without unnecessarily depleting our finite natural resources and further degrading the environment upon which we all depend? and second: can humanity collectively step back from the brink of environmental collapse and, at the same time, lift its poorest members up to the level of basic human health and dignity?[1]

Sustainable development: a definition

This chapter describes the origins of sustainable development: the salient trends and events that mark the development of the concept. It also presents the important models and principles of sustainability that writers and scientists have developed over the last decade. The purpose for this account is to provide the reader with an understanding of how the concept evolved and with a working knowledge of the efforts to make the concepts of sustainability operational.

The Brundtland Commission report, *Our Common Future*, contains the well-accepted and often quoted definition of sustainable development: development that "*meets the needs of the present without compromising the ability of future generations to meet their own needs* (emphasis added)."[2] The definition, certainly by design, is very compelling, hinting at the potential consequences of our current method of economic development. That is, if we continue on our present course—consuming resources and exhausting ecological carrying capacity—we will not leave enough resources for our children (and their children as well) to reach the same quality of life that we enjoy today. By itself, however, this definition does not offer sufficient guidance about what needs to be fixed or how to make the repairs.

The Brundtland Commission definition is also unsettling. For many of us, it is difficult to accept that such a bleak future is possible, particularly for those of us living in the United States. At no other time in history have the people in this country enjoyed more prosperity or a higher quality of life. Not only does this country have the strongest economy in the world, but it also acts as the primary engine for global growth, accounting for almost half of the growth in world demand in 1998.[3] Even today as the world moves out of a recession, the United States economy is leading the recovery.[4] As a result, most citizens of the United States today enjoy an incredible array of affordable goods and services, a situation to which other nations aspire.

Engineers may also find the prognosis of the Brundtland Commission particularly difficult to accept. The growth and prosperity in the United States are due in no small way to the contributions of engineers and scientists working to improve the built environment. "*Engineers are probably the single most indispensable group needed for maintaining and expanding the world's economic well-being and its standard of living* (emphasis added)," notes Richard Weingardt in his book, *Forks in the Road*.[5] He goes on to describe how much of what we have today is made possible because of the work of engineers.

> *If no engineering minds existed, we would still travel at horse-speed, carry water by buckets ... live and work in unlighted spaces without air-conditioning or central heat. ... There would be no high-rise buildings, long span bridges, interstate highway or rail systems, water control dams. ... More people today would be dying young from contaminated water, spoiled food, and unsanitary conditions than of old age, or cancer or heart disease.*[6]

Our remarkably strong economy and quality of life have been built upon a complex set of technological, industrial, and municipal infrastructures that are, perhaps, the most productive in the world. Because of the work of engineers, we find and extract raw materials more cheaply than ever before. We also move these materials efficiently through many interconnected modes of transportation (air, water, and surface) to efficient and productive manufacturing facilities at home and abroad. There they are converted and returned as parts or finished goods to be sold and used by consumers here and abroad. Technological advances and their corresponding engineering applications have led to continuous improvements in the form of better-performing materials; more-efficient extraction methods; and new, more effective production techniques.

At the same time, we have also learned about the detrimental effects of unwanted by-products on the environment and have made a considerable investment to eliminate, treat, or properly dispose of these materials. Yet, in the face of these advances and general prosperity, society is beginning to question whether or not our quality of life is sustainable—for ourselves or for succeeding generations.

In citing his reaction to accepting an environmental award for a company he represented, Paul Hawken in *The Ecology of Commerce* recounts his awakening about the futility of corporate environmentalism.

> *Despite all this good work, we still must face a sobering fact. If every company on the planet were to adopt the best environmental practices of the "leading" companies—say, Ben & Jerry's, Patagonia or 3M—the world would still be moving toward sure degradation and collapse. So if a tiny fraction of the world's most intelligent managers cannot model a sustainable world, then environmentalism as currently practiced by business today, laudable as it may be, is only a part of an overall solution. Rather than a management problem, we have a design problem, a flaw that runs through all business.*[7]

Is our quality of life sustainable?

Our quality of life is derived from what Paul Hawken calls a linear "take-make-waste" production-consumption model of the industrial age, one that draws freely upon energy and raw materials to produce and deliver goods and services to meet consumer needs. As a result, people in this country consume more resources, in terms of both size and per capita, than any other country in the world. Our model of production and consumption is based on the unstated assumption that the earth's carrying capacity (supply of natural resources and ability to assimilate wastes and ecological services) is more than sufficient to support all the activities of a large and increasing population. If it isn't, then we will simply repair the damage, find other sources, or invent substitutes.

> *While there is nothing inherently wrong with a population, even a large one, meeting its needs by consuming resources and creating wastes ... problems do arise, however, when the numbers of people and the scale, composition, and pattern of their consumption and waste combine to have negative effects on the environment, economy and society.*[8]

As noted in chapter 1, evidence of nonsustainable development is starting to appear in many places and on many scales. The reason for this recent phenomenon is that it was only at the end of the 20th century that human activities reached the point at which they could effect significant changes on the environment. Prior to that time, the effects were negligible or confined to small areas. Today many of these effects are huge, freely crossing national borders. The metaphor of the *tragedy of the commons*[9] extends regionally, as in depletion of fishing resources, or globally, as exemplified by ozone depletion or global climate change.

Production-consumption model

To understand the concept of sustainability, we must understand how our system of production and consumption works. Specifically, we must understand the processes in which we

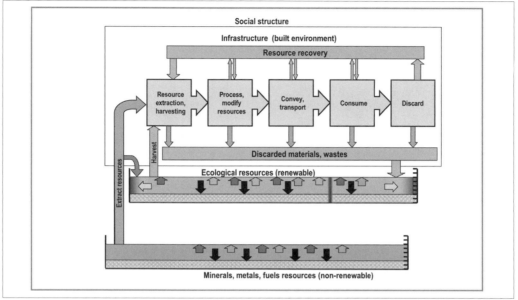

Figure 2-1: Production-consumption model. Source: Wallace Futures Group, LLC.

harvest or extract resources, produce and deliver the products and services, and dispose of unwanted by-products, all of which contribute to our quality of life. A simplified model[10] is presented in figure 2-1.

In the model, we partitioned the resources into two types: renewable and nonrenewable. Renewable resources are the living, ecological resources that we can continuously grow, harvest, and convert into goods and services. As the size of the rectangle in the diagram indicates, the quantities of renewable resources at any given time are finite. However, they are continuously renewed, a function of the health of the ecosystem. Some renewable resources are not economically retrievable, although advances in harvesting techniques could change that. These quantities will remain stable so long as we do not overharvest or damage the ecosystems. Wastes from our systems of extraction, harvesting, production, and consumption are often im-

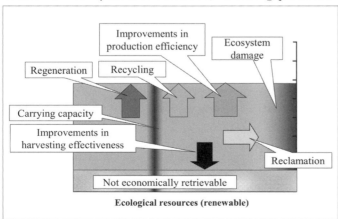

Figure 2-2: Ecological (renewable) resources, a closer look.
Source: Wallace Futures Group, LLC.

properly handled and continue to cause damage to our renewable resources. If allowed to continue, the damage will limit the quantities of renewable resources available. Worse, it may reduce these resources to such a degree as to exceed the carrying capacity of the resource, that is, its ability to renew itself in the face of continuous harvesting and damage to the ecosystem in which it resides.

Minerals, metals, and fuel resources are classified as nonrenewable. These resources are also finite, and some substantial quantity is currently not economically retrievable. In the production-consumption model, nonrenewables are also extracted and converted into goods and services. Some waste materials created during the extraction process may be deposited in the environment, potentially causing damage to the renewable resources.

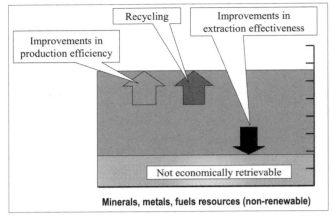

Figure 2-3: **Nonrenewable resources, a closer look. Source: Wallace Futures Group, LLC.**

The current production-consumption process is depicted as five steps: resource extraction or harvesting, processing or modification, transportation, use, and discarding. Unwanted by-products may be created in each step, some of which may reach and cause damage to our renewable resources. Other by-products may be recycled at each step. Finally, all of the production-consumption steps are incorporated in our built infrastructure. That infrastructure, along with the renewables and nonrenewables that are contained within, are ultimately influenced by society.

Conditions for sustainability

Herman Daly gives three conditions of a sustainable society:

1. Rates of use of renewable resources do not exceed their rates of regeneration.
2. Rates of use of nonrenewable resources do not exceed the rate at which sustainable renewable substitutes are developed.
3. Rates of pollution emission do not exceed the assimilative capacity of the environment.[11]

In consideration of social equity aspects of sustainable development, it is important to add this condition:

4. Fair and efficient use of resources that meets human needs.

The first four conditions parallel the four system conditions of *The Natural Step* framework, discussed later in this chapter. Finally, from a practical standpoint, two more conditions must be met:

5. Companies need to stay in business in order to effect and finance the necessary changes.
6. Changes must be accomplished without diminishing substantially our quality of life.

Unfortunately today, the arguments over sustainability have degenerated into two camps. At one extreme are the "apocalyptics" who believe that future generations won't survive unless we drastically reduce resource consumption now, even at the expense of our economy and quality of life. At the other extreme are the "technology-optimists" who dismiss the resource and carrying capacity problems as being untrue or overstated. They believe that any shortages or ecological damage that may arise will be solved; that is, the requisite technologies will be developed in time to "save the day."

Neither position is helpful. An unquestioning reliance on technology to provide "just-in-time" solutions ignores the evidence of a coming unprecedented increase in world population (a fivefold increase between 1950 and 2050) and an attendant increase in resource demand.

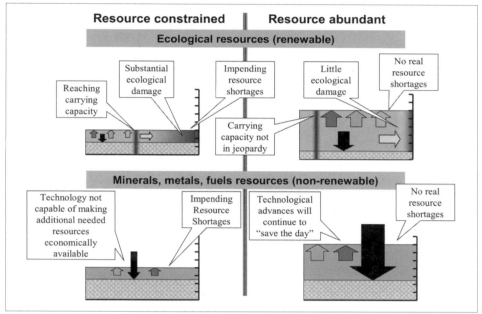

Figure 2-4: Polar polemics: the debate over sustainability. Source: Wallace Futures Group, LLC

Furthermore, that demand will be exacerbated by the aspirations of the developing world to improve its quality of life through increased production and consumption. Unless something is done, these demands will be met primarily through nonsustainable technology. On the opposite side, the remedies offered by the apocalyptics are bleak, calling for large reductions in the use of resources through drastic production and consumption cutbacks. This approach offers a grim near-term future – a serious reduction in our quality of life.

The origins of sustainable development

How did we get to this situation? When did we begin to understand the full ramifications of our system of development? The concept of sustainable development per se did not appear until the late 1980s. However, by most accounts, Rachel Carson is credited with inaugurating the environmental movement through her 1962 publication *Silent Spring*. Writing about the dangers of DDT and other pesticides, Carson detailed in layman's terms how these chemicals entered the food chain and accumulated in animal and human tissue. This exposure, she explained, could cause cancer and genetic damage.

Not surprisingly, the chemical industry was outraged. They attacked Carson as a fanatic, stating that her work was full of errors and warning that, without pesticides, society would be overrun with insects and disease. Industry executives were engaged to write and speak out against her. One chemical company quickly responded by sending to the media a commissioned publication called *A Desolate Year*, a satire of Carson's work in which crops are destroyed by hordes of ravaging insects as pesticide use is reduced.

Although the concept of sustainable development was still 25 years away, Carson's book raised not only the issue of chemical dangers, but of corporate transparency and responsibility.

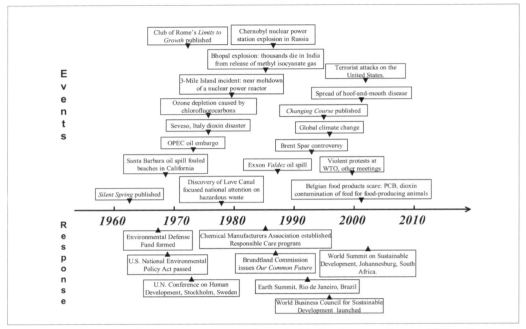

Figure 2-5: Salient events in the movement toward sustainable development. Source: Wallace Futures Group, LLC.

It also brought into question corporate ethics, particularly undue corporate influence over the scientific research of the day.

Carson also challenged what had become a blind faith in what science and technology had to offer. In a 1962 speech to the National Women's Press Club she noted that, for questions about the effects of pesticides on human health, the American Medical Association was referring its members to a pesticide trade association.

> *Now I am sure physicians have a need for greater information on this subject. But I would like to see them referred to authoritative scientific or medical literature and not to a trade organization whose interest it is to promote the sale of pesticides.*[12]

Rachel Carson died in 1964, less than two years after the publication of Silent Spring. But soon after her work was published in 1962, President John F. Kennedy asked his Science Advisory Committee to investigate the issues raised. The committee's report agreed with virtually all of Carson's assertions, and DDT and other pesticides were eventually banned from use.[13] *Time* magazine lists her as one of the top 100 most influential people of the 20th century, citing this work as the "cornerstone of the new environmentalism."[14] *Life* magazine lists the publication of *Silent Spring* as number 70 of the top 100 events of the millennium.

The Limits to Growth: predicting the apocalypse

The idea that society's economic development was pushing the earth's resource and ecological limits has been around for some time. Over the past decades, there have been a number of writings predicting a serious depletion of critical resources or ecological ruin. Perhaps the most recent and famous of the "doomsdayish" predictions was the 1972 report from a group known as the Club of Rome called *The Limits to Growth*. Using sophisticated computer models, this analysis predicted the times when our society would run out of critical resources. This apocalyptic work sold over 9 million copies printed in 30 different languages and sparked outrage and heated debates everywhere. Fortunately, most of its predictions were wrong.

Unfortunately, its errors have been held up as evidence of the folly of making such predictions. Furthermore, it reinforced the notion that our technological cleverness would continue to save the day, providing us with new sources or alternatives to meet our resource and energy needs.

Although the work was criticized as being neo-Malthusian, the warnings raised by the authors soon were to appear prophetic. In 1973, a year after publication, the world was in the throes of an oil embargo organized by members of the Organization of Arab Petroleum Exporting Countries (OPEC). Although these oil shortages were imposed, they delivered a painful lesson about the effects of critical resource shortages and resulted in a reassessment of the United States' strategic position on energy.

United Nations Conference on the Human Environment

In 1972, representatives from 113 industrialized and developing nations met in Stockholm, Sweden, for the first world conference on the environment. Convened by the United Nations (UN), this first-time world conference focused on environmental degradation and "transboundary" pollution—pollution that does not recognize or abide by political or geographic boundaries. The conference produced the Stockholm Declaration and Action Plan which, for the first time, identified environmental problems at a world scale and sought international action to address them. The conference noted the relationship of humans to the natural environment and identified the links between economic and social development and environmental protection.[15]

Many nations responded by passing the first laws for protecting and preserving the environment. The idea of environment reporting, part of the 1969 U.S. National Environmental Policy Act (NEPA), was adopted internationally at the Stockholm Conference.[16]

The Brundtland Commission report

In 1983, the United Nations established the World Commission on Environment and Development and appointed Gro Harlem Brundtland, a former environmental minister and later the prime minister of Norway, to lead it. The charge of the commission was to identify the critical global environmental and developmental issues and to set out a course to address them. Members of the commission included experts from 21 nations, the majority from World 2 and 3 countries. Over the course of three years, the commission sponsored over 75 studies on environment and development issues and held 15 public hearings. In 1987, the "Brundtland" Commission published its report, *Our Common Future*.

The significance of the report was its understanding and recognition that over the last 100 years the relationship between the environment and human endeavors had changed radically. At the beginning of the 20th century, humankind had neither the numbers nor the technological power to significantly affect the world's environment. Now, approaching the close of the 20th century, the population of the planet has increased sharply, moving from 1.7 billion in 1900 to over 6 billion today. More importantly, there has been a corresponding sharp increase in human activities and their power to affect the environment.

> *As the century closes, not only do vastly increased human numbers and their activities have that power, but major, unintended changes are occurring in the atmosphere, in soils, in water, among plants and animals, and in the relationships of all of these. The rate of change is outstripping the ability of scientific disciplines and our current capabilities to assess and advise.*[17]

These trends demand that a different pathway be found for future economic development. Development, the commission said, must be sustainable. All future development must enable society to maintain and improve its quality of life without diminishing the ability of future generations to maintain and improve their quality of life. This is a new and important concept that is changing the way people think about economic development and its effect on the environment and society.

The report called for a new form of development—sustainable development.

> *Humanity has the ability to make development sustainable—to ensure that it meets the needs of the present without compromising the ability of future generations to meet their own needs (emphasis add-ed). The concept does imply limits—not absolute limits but limitations imposed by the present state of technology and social organization on environmental resources and by the ability of the biosphere to absorb the effects of human activities.*[18]

Our Common Future was a call to action for the nations of the world to come together and create goals, strategies, and programs for moving toward a more sustainable future. Since its publication, there have been two world summits, culminating in treaties and action plans, all geared to addressing the problems raised in the report. The commission called for an international conference to be assembled to review progress, promote follow-up, and set benchmarks to measure progress toward sustainability. That conference, called the Earth Summit, was held in 1992 in Rio de Janeiro.

Even before the Brundtland Commission produced its report, the concept of sustainable development had been introduced. In 1980, the Independent Commission on International Development issued a report calling for "sustainability in social development and economic growth, as well as the natural environment." Named the Brandt Commission after its chairman, former German chancellor Willy Brandt, the commission produced a report, *North-South*, noting that *"the conquest of poverty and the promotion of sustainable growth are matters not just for the survival of the poor, but for everyone."*[19]

The 1992 Earth Summit

Responding to the call from the Brundtland Commission report, 30,000 people including the leaders from more than 100 countries came to Rio de Janeiro, Brazil, for this largest-ever world conference on the environment. Officially known as the United Nations Conference on Environment and Development, the intent of the conference was to take a fresh look at economic development in light of the Brundtland Commission report's findings. Delegates to the summit concluded that an entire overhaul of national attitudes and behavior was required in order for the nations of the world to move onto a sustainable pathway.

The summit produced five international agreements, all of which remain a significant influence in national and world agendas:

- *Agenda 21.* A 40-chapter, 800-page agenda for action that calls for a broad set of programs leading to sustainable development. While the agenda was nonbinding, there was an implied commitment by the signatory nations to address the problems and issues raised.

- *Rio Declaration.* A declaration of 27 universal sustainable development principles signed by the delegates.

- *Framework Convention on Climate Change.* An agreement signed by 154 countries to address the emissions of the so-called greenhouse gases and their effects on global climate.

- *Convention on Biodiversity.* An agreement signed by 168 countries for conservation and sustainable use of biodiversity.

- *Statement of Forest Principles.* A legally nonbinding statement that contains 15 principles concerning the sustainable management of forest resources.

Before the Earth Summit was under way, the private sector was responding to the findings of the Brundtland Commission. The Business Charter for Sustainable Development, developed in 1991 by the International Chamber of Commerce, published 16 sustainable development principles endorsed by 600 companies. In addition, the CEOs of 50 major multinational corporations founded the Business Council for Sustainable Development (BCSD). In 1995, the BCSD merged with the World Industry Council for the Environment and became the World Business Council for Sustainable Development (WBCSD). Today WBCSD has over 165 members, including many of the largest companies in the world.

Agenda 21

Perhaps the most important outcome of the 1992 Earth Summit was Agenda 21, a strategy for moving toward sustainable development in the next century. It was developed as a blueprint for action for virtually every aspect of human activity. The report contains 40 separate sections of concern and outlines 120 separate action programs.

The importance of Agenda 21 is that it set out, for the first time, a comprehensive set of goals and priorities for the major resource, environmental, social, legal, financial, and institutional issues. For each issue, the document lays out a plan of action along with a program and an estimate of the financing needed to run it. While not a legally binding document, it was adopted by nations representing 98 percent of the world's population.

Thankfully, the framers of Agenda 21 recognized that the systematic acquisition of sound, credible information about current resource, ecological, and social conditions was critical in the movement toward sustainable development. Without such information, policymakers will be hard pressed as to how to set strategies and priorities and how best to spend scarce resources.

Chapter 35 in Agenda 21, "Science for Sustainable Development," focuses on the role of science in the management and preservation of the environment, noting that a better understanding of the land, oceans, and the air *"is essential if more accurate estimates are to be provided concerning the carrying capacity of planet Earth and of its resilience to the stresses placed upon it by human activities* (emphasis added)."[20] This section of the document calls for the preparation of an inventory of the natural and social science data relevant to sustainable development and the identification of research needs and priorities. Chapter 40, "Information for Decision-making," points to the gap that exists "in the availability, quality, coherence, standardization and accessibility of data ... which has seriously impaired the capacities of countries to make informed decisions concerning environment and development."[21]

Integrated ecosystem assessments

In response to this need, 16 international organizations began a $4 million study to begin an integrated assessment of the condition of global ecosystems and to identify information gaps. This study, the Pilot Analysis of Global Ecosystems (PAGE), analyzed five ecosystems[22] at the continental and global scales in terms of the goods and services these ecosystems produce. All five ecosystems are showing signs of degradation.[23]

The PAGE study has led the way for a more comprehensive study, the Millennium Ecosystem Assessment. The assessment is a four-year, $20 million effort begun in April 2001 to provide decision makers and the public with sound, peer-reviewed information about the "condition of ecosystems, consequences of ecosystem change, and options for response."[24] Sponsored by the Global Environment Facility, the United Nations Foundation, the David and Lucile

Packard Foundation, and the World Bank, it is the first comprehensive attempt to develop a scientifically defensible understanding of the state of the world's ecosystems in terms of their ability to deliver goods and services now and into the future. This effort will assess the conditions and changes in ecosystems at multiple scales, from the global level reaching down into regions, nations, river basins, and local villages. The assessment reports are scheduled for release in early 2005. In addition to assessing ecosystem conditions, the work will also identify important scientific uncertainties and information gaps.

The importance of the Millennium Ecosystem Assessment is that it offers a credible and comprehensive alternative to the current profusion of ecological indicators and assessments now being developed and promoted by various groups, often in support of specific agendas and points of view. Although well meaning in their intentions, the publication and dissemination by these groups of literally hundreds of so-called sustainable development indicators only serve to confuse rather than inform the public about the condition of ecological systems and critical resources.

The World Summit on Sustainable Development

In September 2002, 22,000 people came to the World Summit on Sustainable Development in Johannesburg, South Africa, to reflect on what had been accomplished in the 10 years since the Rio conference. Despite the sweeping strategies, plans, and political commitments of Agenda 21, the consensus going into this summit was that little progress had been made and, in fact, environmental and social conditions had worsened over the last 10 years.

The tenor of this summit was implementation. Recognizing that more political debates and philosophizing would not cure the worsening environmental and social problems, the participants identified known critical problems and issues and set out a comprehensive set of targets, timetables, and commitments for improvement. The participants understood that there were no miracle cures available. Solutions would come only through practical and sustained efforts. Some of the key goals and targets included the following:

- Halve, by the year 2015, the proportion of people without access to safe drinking water and basic sanitation.

- Aim, by 2020, to use and produce chemicals in ways that do not lead to significant adverse effects on human health and the environment.

- Achieve, by 2010, a significant reduction in the current rate of loss of biological diversity.

- Encourage the application, by 2010, of the ecosystem approach for the sustainable development of the oceans.

- On an urgent basis and where possible, by 2015, maintain and restore depleted fish stocks to levels that can produce the maximum sustainable yield.[23]

One important characteristic of these goals and targets is that they are real and recognized problems that, if solved, will make a measurable contribution to sustainable development. Another is that they are achievable at reasonable costs and within a reasonable period of time, bringing hope that the summit 10 years from now will be able to show real progress. The test over the next decade will be the extent to which these commitments are met.

MILLENNIUM DEVELOPMENT GOALS TO BE ACHIEVED BY 2015	
HALVE EXTREME POVERTY AND HUNGER 1.2 billion people still live on less than $1 a day. But 43 countries with more than 60 percent of the world's people have already met or are on track to meet the goal of cutting hunger in half by 2015.	**REDUCE MATERNAL MORALITY BY THREE QUARTERS** In the developing world, the risk of dying in childbirth is one in 48. But virtually all countries now have safe motherhood programmes and are poised for progress.
ACHIEVE UNIVERSAL PRIMARY EDUCATION 113 million children do not attend school, but this goal is within reach; India, for example, should have 95 percent of its children in school by 2005.	**REVERSE THE SPREAD OF DISEASES, ESPECIALLY HIV/AIDS AND MALARIA** Killer diseases have erased a generation of development gains. Countries like Brazil, Senegal, Thailand and Uganda have shown that we can stop HIV in its tracks.
EMPOWER WOMEN AND PROMOTE EQUALITY BETWEEN WOMEN AND MEN Two-thirds of the world's illiterates are women, and 80 per cent of its refugees are women and children. Since the 1997 Micro-credit Summit, progress has been made in reaching and empowering poor women, nearly 19 million in 2000 alone.	**ENSURE ENVIRONMENTAL SUSTAINABILITY** More than one billion people still lack access to safe drinking water; however during the 1990s, nearly one billion people gained access to safe water as many to sanitation.
REDUCE UNDER-FIVE MORTALITY BY TWO THIRDS 11 million young children die every year, but that number is down from 15 million in 1980.	**CREATE A GLOBAL PARTNERSHIP FOR DEVELOPMENT, WITH TARGETS FOR AID, TRADE AND DEBT RELIEF** Too many developing countries are spending more on debt service than on social services. New aid commitments made in the first half of 2002 alone, though, will reach an additional $12 billion per year by 2006.

Figure 2-6: Millennium Development Goals. Source: Indicators for Monitoring the Millennium Development Goals, (New York, United Nations Development Group, 2003).

Sustainable development principles and concepts

Over the past 15 or so years since the issues of sustainable development were raised, industry, government, and nongovernment organizations have been attempting in various ways to make sustainable development operational. What have emerged are a number of sustainable development principles and concepts, each attempting to clarify and guide the transition to sustainable development. If sustainability is a goal, then it follows that coherent strategies and policies are necessary to help organizations reach that goal, by setting down standards, assumptions, ethics, and values to guide the organizations in their work.

In that time period, a number of important sustainable development concepts and principles have emerged, many of which have become part of the sustainable development lexicon. When discussing sustainable development with clients, it is important for engineers to understand not only the concepts and principles themselves, but their origins as well.

Clients who have made organizational commitments to sustainable development have done so after exhaustive research into its origins, history, and relevant issues. They will be quite conversant with the principles and concepts of sustainable development. Therefore, it is essential that the engineer become familiar with these principles and concepts.

Table 2-1 summarizes the key principles and concepts of sustainable development. They are discussed in detail later in the chapter.

The triple bottom line

Business often discusses sustainability performance in terms of a triple bottom line: measures that go beyond financial performance into social and environmental performance. Proposed

Table 2-1: Examples of sustainable development principles and concepts

Name	Origin	Description
Bellagio Principles	International Institute for Sustainable Development (IISD). Set forth at a November 1996 meeting at the Rockefeller Foundation's Study and Conference Center in Bellagio, Italy, of an international group of measurement practitioners and researchers who reviewed progress on sustainable development.	The Bellagio Principles consider the four aspects of assessing progress toward sustainable development: (1) the establishment of a vision of sustainable development and clear goals, (2) the content of the assessment and the need to merge a sense of the overall system with a practical focus on current priority issues, (3) the key issues of the process of assessment, and (4) the necessity for establishing a continuing capacity for assessment.
Caux Round Table Principles for Business	The Caux Round Table www.cauxroundtable.org	These are seven principles based on two ethical ideals: kyosei and human dignity. Kyosei is a Japanese concept for living and working together for the common good, enabling cooperation and mutual prosperity to coexist with healthy and fair competition.
CERES Principles	CERES www.ceres.org	The CERES Principles consist of a 10-point code of environmental conduct. They were originally known as the Valdez Principles, developed in 1989 after the Exxon Valdez oil spill. Currently, they are endorsed by 70 companies.
Enlibra	Western Governors' Association www.westgov.org	Enlibra is a set of principles for protecting air, land, and water that have proven effective in resolving environmental and natural resource disputes. The word Enlibra was coined by the Western Governors' Association to symbolize balance and stewardship.
The Earth Charter	The Earth Charter Initiative www.earthcharter.org	This is an initiative developed in 1994 by Maurice Strong, the secretary general of the Earth Summit and chair of the Earth Council, and Mikhail Gorbachev, president of Green Cross International This charter is a declaration of principles for building a just, sustainable, and peaceful global society in the 21st century. The drafting of an earth charter was part of the unfinished business of the 1992 Rio Earth Summit.
Hannover Principles	Developed for the Hannover, Germany, 2000 World's Fair http://www.mcdonough.com/principles.pdf	In 1991, the city of Hannover, Germany, asked William McDonough to write out the general principles of sustainability for the 2000 World's Fair. The nine Hannover Principles have become the foundation for sustainable design.
Precautionary Principle	Emerging principle included in the Rio Declaration from the 1992 UN Conference on Environment and Development, also known as Agenda 21	In order to protect the environment, the precautionary approach shall be widely applied by nation-states according to their capabilities. Where there are threats of serious or irreversible damage, lack of full scientific certainty shall not be used as a reason for postponing cost-effective measures to prevent environmental degradation.
Three pillars of sustainability	World Business Council for Sustainable Development (WBCSD) www.wbcsd.org	This concept follows the triple bottom line approach, but uses a building metaphor, with economic, environmental, and social issues as the "pillars."
Triple bottom line	John Elkington, SustainAbility www.sustainability.com	In its basic form, the triple bottom line is a framework for measuring corporate performance: economic, environmental, and societal. Taken more broadly, it can be used to describe the dynamics of shifting forces: economic, environmental, and social. It uses the metaphor of tectonic plates moving against each other. "Shear zones" between plates create "tremors," i.e., issues and conflicts.
Wingspread Consensus Statement	Meeting of scientists, philosophers, lawyers, and environmental activists on January 26, 1998, at Wingspread, headquarters of the Johnson Foundation www.sehn.org/wing.html	Reached agreement on the necessity of the Precautionary Principle in public health and environmental decision making. The Wingspread Consensus Statement on the Precautionary Principle.

by John Elkington of SustainAbility, the term *triple bottom line* addresses sustainability in terms of the business metaphor—return on capital investment. In the case of sustainable development, capital takes three forms. Financial capital, the traditional business metric, is included. However, there are two other forms of capital—natural capital and social capital—both of which can deliver returns against funds expended.

Natural capital recognizes the services provided by natural resources and looks at returns gained by using less raw materials and creating less waste. In his book *Cannibals with Forks*,[26] Elkington identifies two main forms of natural capital: critical natural capital and renewable, replaceable, or substitutable natural capital. *"The first form embraces natural capital which is essential to the maintenance of life and ecosystem integrity; the second forms of natural capital which can be renewed (e.g., through breeding or relocation of sensitive ecosystems), repaired (e.g., environmental remediation or desert reclamation), or substituted or replaced (e.g., growing use of man-made substitutes, such as solar panels in place of limited fossil fuels)"*[27]

Social capital recognizes the services provided by employees as well as the social infrastructure in which the organization resides. Here, returns come in the form of such things as stakeholder trust and employee morale. For investors, triple bottom line performance is a better measure of the long-term success of a corporation than financial performance alone, capturing the breadth of sustainable development.

Elkington moves to another metaphor in describing how the components that make up the triple bottom line interact. According to Elkington, the three elements—economic, environmental, and social—are in constant movement, according to the ebb and flow of issues contained therein. Their

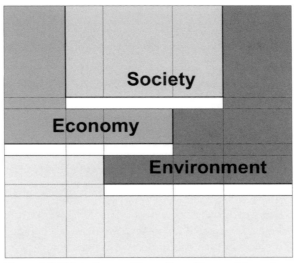

Figure 2-7: Triple Bottom Line – Shear Zones. Adapted from John E. Elkington, Cannibals Without Forks: The Triple Bottom Line of 21st Century Business, New Society Publishers, Gabriola Island, BC, Canada 1998. pp. 69-96.

interactions with one another are much like tectonic plates, creating disruptions at the shear zones. For example, issues concerning the sustainable use of resources for economic growth are formed at the economic-environmental zone. Issues concerning environmental justice appear at the social-environmental zone. Businesses concerned with their triple bottom line performance should explore carefully the issues at these zones that are important to their stakeholders.[28]

The three pillars of sustainability

The World Business Council for Sustainable Development[29] uses a building metaphor to describe its model for sustainable development (see figure 2.8). Following the precepts of the triple bottom line, the WBCSD sees sustainability as resting on three pillars: economic growth, ecological balance, and social progress. Concepts such as eco-efficiency, corporate social responsibility (CSR), and innovation are part of the support structure, as are framework conditions and the financial markets. Framework conditions are the government operations, policies, and regulatory frameworks within which businesses must operate. Although not as elegant as other models, the WBCSD model is important in that it depicts the way in which many of the world's leading corporations think about the structure of, and the issues surrounding, sustainable development. For the most part, the WBCSD membership believes that the means for achieving sustainability are through eco-efficiency: adding value to goods and services while using less resources and producing less waste.[30]

Making sustainability operational

In addition to sustainability principles, there are a number of important frameworks: models for thinking about the issues of sustainable development and developing ways to manage organizations.

The Natural Step framework

The Natural Step (TNS) framework is a concept that converts the complexity of environmental problems into a set of basic principles, presented as four system conditions for sustainability.

> *In order for society to be sustainable, nature's functions and diversity are not systematically:*
>
> *1. subject to increasing concentrations of substances extracted from the Earth's crust;*
>
> *2. subject to increasing concentrations of substances produced by society; or*
>
> *3. impoverished by overharvesting or other forms of ecosystem manipulation.*
>
> *And,*
>
> *4. resources are used fairly and efficiently in order to meet basic human needs worldwide.*[31]

The Natural Step is the United States affiliate of a not-for-profit Swedish organization Det Naturliga Steget, founded in 1989 by physician Karl-Henrik Robèrt. Its mission is to guide business and government onto a sustainable path through research and education. Robèrt, an oncologist and cancer researcher, noticed an increase in the number of leukemia cases in children and traced the cause back to increased exposures to toxic materials in the environment. He also observed that much of the environmental debate was all about the effects and the timing of a myriad of environmental problems, not the underlying causes. He characterized this debate as monkeys chattering about the withering leaves of a dying tree. Is the prospect of global warming really a threat? Are persistent organic pollutants harmful? When should we convert to nonrenewable energy sources? While these questions are interesting, they obscure the real issue.

Robèrt and his colleagues reasoned that while all of these environmental questions are subject to debate, the underlying causes are not. They worked to find a common ground— a set of principles that could be used to make decisions about the environment and society. With the help of 50 scientists, Robèrt developed a consensus document that described how the biosphere functions and interacts with society's activities. After 21 iterations, the document was published and distributed to all the schools and households in Sweden.

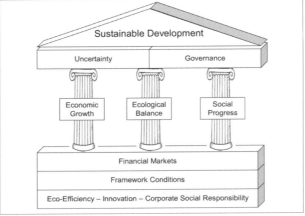

Figure 2-8: The Three Pillars of Sustainability. Adapted from the World Business Council for Sustainable Development: Three Pillars of Sustainability. Reference: WBCSD Strategy 2010, 12 June 2001.

Later Robèrt worked with physicist John Holmberg to produce the four system conditions for sustainability. These system conditions, together with the consensus document, form the basis for The Natural Step framework.

The importance of The Natural Step framework is that it recognizes the futility of debating a seemingly endless array of environmental effects and interactions. Instead, it moves the debate to a set of fundamental principles on which all can agree. For example, instead of debating what levels of polychlorinated biphenyls (PCBs) the human body can safely absorb, we can all agree that it is not a good idea to allow PCBs to enter the environment in the first place. At this writing, more than 70 municipalities and 60 companies, including some of the largest in the world, have incorporated the Natural Step into their operations.

The Natural Step metaphor: the funnel

The TNS resources funnel is shown in figure 2.9, depicting the overall situation for the earth. Resource demands are increasing rapidly, following the environmental impact equation popularized by Paul and Anne Ehrlich:

I (environmental impact) = P (population) · A (per capita affluence) · T (technology)

T, the technology term, is an index of environmental damage done by the technologies used to support human consumption.[32] Fueled by a growing population with increasing affluence, and using linear (take-make-waste) technologies, society is drawing upon a finite supply of life-sustaining resources. As a result, the gap between resource supply and demand continues to decrease, reducing society's margin for taking any counteraction.

In their book *The Natural Step for Business*, Brian Nattrass and Mary Altomare explain the TNS resources funnel.

> *Imagine the walls of a giant funnel. The upper wall is resource availability and the ability of the ecosystem to continue to provide services. The lower wall is societal demand for resources that are converted into goods and services such as clothes, shelter, food, transportation, etc., and ecosystem services such as clean water, clean air, and healthy soil. As aggregate societal demand increases and the capacity to meet those demands decreases, society as a whole moves into a narrower portion of the funnel.*
>
> *As the funnel narrows, there is less room to maneuver and there are fewer options available. The inactive company that remains oblivious to the changing environmental realities is likely to hit the wall and go out of business. The reactive company waits until it gets clear signals from the environment, often by running into the wall, and then it must react quickly or fail.*[33]

Ray Anderson, in his book *Mid-Course Correction*, proposes a further refinement to the model, which addresses the design concerns expressed by William "Bill" McDonough.[34] Anderson revisits the impact equation, which describes the lower wall of the curve.

Anderson observes that the technology term applicable to the Natural Step model should be depicted as T_1, referring to the extractive, linear (take-make-waste), fossil fuel-driven, abusive, consuming, and otherwise unsustainable technologies of what he calls the first industrial age. He then proposes a new equation that identifies T_2 technologies. These technologies are renewable, cyclical, benign, and focused on resource productivity. The equation can be restated to read:

I (environmental impact) = P (population) · A (affluence per capita) / T_2 (benign technologies)

In this case, the T_2 term is the index of environmental and resource productivity improvement. Instead of technology being cast as the cause of a degraded future, technology is a solution, offering an improved quality of life.

Making The Natural Step framework operational

For companies that want to not only avoid hitting the wall but set their course toward a sustainable future, The Natural Step framework poses a four-step process: the A-B-C-D Analytical Approach. These four steps are described below and are depicted in figure 2.10:

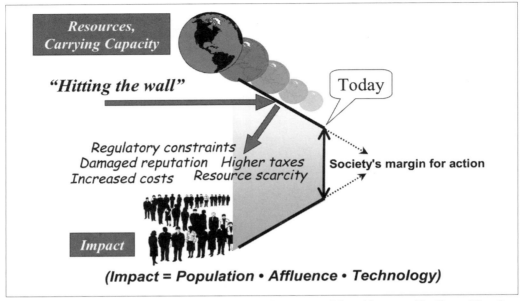

Figure 2-9: The Natural Step funnel. Adapted from Brian Natrass, Mary Altomare, *The Natural Step for Business*, New Society Publishers, Gabriola Island, BC, Canada, 1999.

- *Awareness.* Before anything can be accomplished, the organization needs to recognize and agree that the current societal system in which it operates is not sustainable. Overall, the current economic, environmental, and social conditions and trends cannot be maintained in the long term, and the range of options for taking action is narrowing.

- *Baseline mapping.* Based on this awareness, the organization must assess its current operations—material and energy flows, social policies and conduct, ecological impact—in terms of the four system conditions.

- *Clear vision.* The organization needs to develop a vision of what it will look like in a sustainable society.

- *Down to action.* Use backcasting to determine what steps need to be taken to achieve the vision and then set priorities. In backcasting, the organization sets its sights on the vision of a sustainable future created in the previous step. Then, looking back from the vision to its current baseline of operations, it determines the strategies, plans, and programs necessary to achieve that vision and the metrics to measure progress.[35]

In the action step, strategies, plans, programs, and metrics are continually reviewed and modified based on the results achieved as well as new experiences and circumstances. Here, the four system conditions act as a compass guiding the organization toward its vision of sustainability.

Commitments to and actions by organizations to achieve this vision of sustainability lead them to making changes in their processes, products, and services, following the precepts of the four system conditions. These changes will include such things as reducing the use of nonrenewable resources such as petroleum products and nonrecyclable minerals and metals, a stoppage of using nonbiodegradable materials, conserving natural resources, and restoring the carrying capacity of the environment. Many of these actions are embedded in the concept of natural capitalism.

Natural capitalism

In their book *Natural Capitalism*, authors Paul Hawken, Amory Lovins, and L. Hunter Lovins make operational John Elkington's triple bottom line concept. When economists discuss capital, they usually mean financial capital: monies invested to produce commercial goods and services, and a return on the investment. But there is another branch of economists—ecological economists—who recognize three other types of capital: human capital, manufactured capital, and natural capital. Human capital consists of the aggregate knowledge, skills, and competencies of individuals within an organization or institution. Manufactured capital refers to

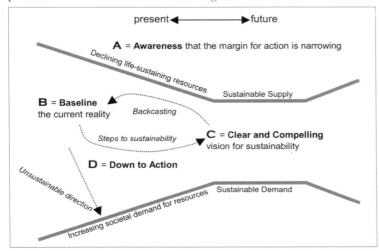

the physical aspects of capital, that is, the factories, buildings, dams, roads, ports, and other humanmade improvements. Natural capital refers to the nonrenewable resources (minerals, metals, and fuels) and the renewable (ecosystems). Nonrenewable resources supply many of the essential materials. Industry uses the first three to transform the fourth into the products and services of our daily lives.

Figure 2-10: Implementing the Natural Step: the A-B-C-D process. Adapted from the Natural Step Canada, "A-B-C-D Process," http://www.naturalstep.ca/implementation.html.

Throughout the Industrial Revolution of the last 150 years, we easily recognized the value of financial and manufactured capital. Of late, human capital is becoming more recognized for its intrinsic value as we become a society of knowledge workers. Not so for natural capital. Even though it provides an extraordinary array of ecosystem services, natural capital is valued as a source of exploitable resources and treated as if its supplies were inexhaustible. Consequently, natural capital is being degraded and liquidated by the wasteful use of such resources of energy, materials, water, fiber, and topsoil.

> *With a dangerously narrow focus, our industries look only at the exploitable resources of the earth's ecosystems—its oceans, forests and plains—and not at the larger services that these ecosystems supply for free. ... Forests, for instance, not only produce the resource of wood fiber but also provide such ecosystem services as water storage, habitat, and regulation of the atmosphere and climate.* [36]

The total annual value of ecosystem services is conservatively estimated at $33 trillion, a figure close to the world's total annual gross product.[37] That valuation assumes that substitutes for these services can be found or that the services can be accomplished by human-engineered systems. For most of these services, most of which are essential for life, there are no known substitutes, and they are therefore priceless.

In their book, authors Paul Hawken, Amory Lovins, and L. Hunter Lovins pose a pivotal question: What if natural capital were treated like the other forms of capital—used productively with the profits reinvested to maintain and enhance the resource base?[38] The authors point out that at the start of the Industrial Revolution, natural resources were abundant and labor was the limiting factor of production. Back then, it made sense for society to focus its efforts on making people more productive. Today the situation is reversed. Now there is an abundance of people and an increasing decline in natural resources and ecosystems. Why not respond to this changing pattern of scarcity by increasing the productivity of, and investing in, natural capital?

The authors offer the concept of *natural capitalism*, a new approach to the way natural capital is valued and treated by business. Natural capitalism has four principles:

- *Dramatically increase the productivity of natural resources.* The authors show that, by making changes in production design and technology, companies are developing ways to up the productivity of natural resources by five, 10, or even 100 times. The resulting savings in operational costs, capital investment, and time can increase profits and help implement the other three principles.

  For example, by refurbishing the glazing on a 20-year-old building in Chicago, Illinois, with new window glazing technology, it is possible to reduce the cooling load from 750 tons of air-conditioning to less than 200 tons. The new window technology lets in six times the light but a tenth less heat. Better use of natural light would allow for a reduction in lighting systems and the corresponding heat generation. Savings in air-conditioning and energy costs would more than make up for the increase in glazing costs over conventional glazing.[39]

- *Shift to biologically inspired production models.* Natural capitalism seeks to eliminate the entire concept of waste by following nature's models, replacing the current take-make-waste systems with closed loop production systems. Here, every production output is either returned harmlessly to the ecosystem as a nutrient or becomes a feedstock for another production process. This approach is akin to the concepts of industrial ecology, by-product synergy, and biomimicry.

  In her book *Biomimicry: Innovation Inspired by Nature*, author Janine Benyus extends the role of nature from a model for production to a model for ideas. Consider Kevlar, one of the strongest humanmade materials. But to make Kevlar, we mix petrochemicals with sulfuric acid at a high temperature and pressure and then squeeze out fibers. Spiders, on the other hand, make a silk that is as strong but much tougher than Kevlar, but they do it at room temperature and pressure and without the acid. For feedstocks, they use flies and crickets. They don't need to drill for oil.[40]

- *Move to a solutions-based business model.* Many of today's businesses focus on the sale of goods rather than the services those goods provide. Under the conventional business model, growth and profitability are tied directly to the quantities of goods sold, forcing companies to sell more "stuff" to more people. However, under natural capitalism, the focus shifts to delivering value through services. In other words, instead of selling lightbulbs, sell illumination.

  This concept can be applied broadly. For example, Interface, Inc. is moving from selling carpets to leasing floor covering services. CEO Ray Anderson determined that people do not want to own carpeting. They just need floor covering that is comfortable to walk on and looks good. Interface now leases carpet to its customers in the form of carpet tiles. They replace worn carpet—usually just one-fifth the total carpeted area—with new tiles as needed. This approach reduces costs, creates more employment, and delivers a higher profit.

- *Reinvest in natural capital.* The authors call for industry and government to restore, maintain, and expand the world's ecosystems and to discontinue the practice of eroding the basis for future prosperity. We invest in human and manufactured capital. It makes sense to invest in natural capital as well.

The idea of reinvesting in natural capital has led business entrepreneurs to reexamine the conventional wisdom of production. For example, a New Mexico rancher is raising the carrying capacity of his rangelands by changing the way he grazes his cattle. Instead of placing the herd in one large location, the rancher moves them from place to place, allowing them to graze intensively but briefly in one area. This follows the grazing patterns of wild grassland animals—constantly on the move. This idea, called *management-intensive rotational grazing*, increases the carrying capacity of the range and increases profits for the rancher.[41]

Eco-efficiency

In its simplest terms, *eco-efficiency* means doing more with less: delivering better products and services while using less materials and energy to do so. As the term implies, it is a set of measurements, ratios of various forms of input and output used to gauge industry performance.

The term was coined by the Business Council for Sustainable Development.[42] At the 1992 Rio Earth Summit, business was challenged to show its contribution to sustainable development. Business responded with the publication *Changing Course*,[43] written by Stephan Schmidheiny in collaboration with the council. The book presented compelling arguments for how competitively priced goods and services could be delivered while simultaneously reducing ecological impacts and resource use. Since then, business, through its council successor, the World Business Council for Sustainable Development, has spent a great deal of time and energy developing eco-efficiency principles and practices.

Eco-efficiency circumscribes a sustainable development strategy for business—a way for business to make progress toward sustainable development while improving shareholder value. Since its inception, the concept has expanded from a metric for resource productivity to a catalyst for innovation. Its themes and principles cause companies to take a new look at their products and processes. By thinking in eco-efficiency terms, companies are avoiding costs they never realized they had, improving their image, reducing business risk, and creating new products and services that turn out to be better for the customer and better for the environment.

Eco-efficiency has five core themes:[44]

- An emphasis on service. For most companies that sell products, success is measured on the throughput of products: the more products sold, the higher the revenue and the higher the profit. Eco-efficiency proposes that companies rethink what their customers really value. Customers do not want to own a product for its own sake. Rather, they want to receive the service the product provides. By shifting its business model toward service, a company can shift its basis of competition. Instead of being a low-cost commodity supplier of materials, a company can become a value-priced supplier of materials and services, helping clients solve problems.

- *A focus on needs and quality of life.* Eco-efficiency proposes that companies rethink what their customers really need and how the products and services they sell contribute to a better quality of life. This theme asks companies to think about the unmet needs of customers, now and in the future, for example, how will the products and services they deliver improve quality of life? While selling more automobiles and building new highways may satisfy short-term needs, they will not solve future traffic congestion problems and will add to pollution. By thinking ahead, companies may identify new, innovative approaches that can result in a distinctive competitive advantage.

- *Consideration of the entire product life cycle.* Most companies focus only on the products and services they deliver and do not consider the ramifications of their designs or processes up and down the value chain. By considering the impacts of design and process decisions through the entire life cycle, companies can identify numerous ways to improve the value of products to their clients. As suggested in earlier examples, the cost to put up a new building is only a small fraction of the costs to operate it. As Scott Johnson notes,

  > A building owner generally spends nine times more to build a facility than to design it, and thirty times more for O&M than for design, and 460 times more to pay the people that will work in the facility (over its useful life) that was spent on the design.[45]

  Thus a small investment in energy-saving designs and designs that improve the work environment can bring in great benefits over the life of the building.

  Some farsighted companies have taken this theme much farther, by looking outside their industry boundaries. Following the natural capitalism theme of imitating biological models, they are envisioning a closed industrial ecosystem in which the wastes of one company can be used as feedstocks for another. Such a model has been created in the ecoindustrial park of Kalundborg, Denmark. Companies are also teaming up to look for similar exchange opportunities using the by-product synergy process, presented later in this chapter.

- *Recognition of the limits of ecocapacity.* Companies must operate recognizing the limits of the earth's carrying capacity, doing what they can through continuous improvement to reduce the environmental impact of their operations. Here, it is recognized that companies cannot have a precise measurement of their individual impact, especially when carrying capacity varies by time and location. But companies need to recognize that carrying capacity is a limiting factor to their operations. Not paying attention to this factor may cause them to "hit the wall," in The Natural Step terminology.

- *A process view—eco-efficiency—as much a journey as a destination.* In order to achieve sustainability, society must convert from its current take-make-waste production-consumption model to one that meets the principles of sustainable development. As stated earlier, this will be a long and arduous journey, requiring what amounts to a total overhaul of our current production and consumption infrastructure. It will involve the application of sustainable technologies and processes, many of which have yet to be invented. Recognizing that society has to start somewhere, eco-efficiency provides a platform for assessing the current situation and tracking progress toward sustainability.

Eco-efficiency also has seven principles or goals that companies need to take into account in order to improve their operations, products, and services. These can be converted into measurable objectives:

- *Reduce the material intensity of goods and services.* Reduce the amount of materials used per unit of output over the life cycle of the product or service. Opportunities exist not only in reducing the materials used to produce the product, but also in reducing product packaging. Following the theme of service emphasis, companies can look for ways to reduce product usage, while enhancing the services they provide to their customers.

- *Reduce the energy intensity of goods and services.* Reduce the amount of energy used per unit of output over the life cycle of the product or service. Consuming energy not only taxes the supply of nonrenewable resources, but creates a huge amount of waste and pollution. As discussed earlier, much of the energy savings can be accomplished without inventing new technologies. Significant savings can be gained just by taking a fresh look at the problem.

- *Reduce toxic dispersion.* The emission of toxic or otherwise harmful substances into the environment not only can cause negative impacts on public health, but can also reduce ecological carrying capacity. Companies are finding that reducing toxic emissions results in lower costs and lower business risk.

- *Enhance material recyclability.* Instead of sending unused or unwanted by-products off-site for disposal, companies are encouraged to reuse or recycle these materials. The best strategy is to refurbish and reuse the materials or items in a manner that maintains, to the extent possible, its original value.

- *Maximize sustainable use of renewable resources.* Using renewable resources to the extent possible, including finding and employing substitutes for nonrenewables, will help preserve the essential and nonreplaceable materials. However, it is important to ensure that renewable resources are harvested in such a way as to maintain or enhance the renewable asset.

- *Extend product durability.* Designing products that last longer or that can be readily repaired or refurbished and reused saves materials and energy.

- *Increase service intensity of goods and services.* Design products or services in a way that can meet, or can be easily changed to meet, multiple customer needs. Ideas include shared use (leased products or services or shared facilities), multi-functionality (heat pumps that provide heating, air-conditioning, and hot water or transportation coordination that reduces "deadheading"), and upgrading (changing to a modular design to simplify changing the functionality of equipment and systems).[46]

Using these principles, a company can incorporate eco-efficiency into its operations. The principles can be turned into measurable objectives, which then can be monitored, benchmarked, and factored into business plans and strategies.

Eco-effectiveness

Eco-effectiveness is a term coined by Bill McDonough and Michael Braungart to describe their model for achieving sustainability. In their article "The NEXT Industrial Revolution,"[47] the authors argue for a new sustainability construct. Products, they say, are composed of two types of materials: (1) *biological materials* that degrade and become food for the biological cycle or (2) *technical materials* (metals, plastics, solvents, etc.) that are kept in the technical cycle. The authors call both these material types "nutrients" since they become "food" for subsequent products.

Biological nutrients flow in the biological system, or metabolism. These materials are inherently safe for humans and the environment. They are incorporated into what the authors call "products of consumption," products such that, after use, can be returned to the environment where they will become nutrients in the production of other biological materials.

Technical nutrients flow in the technical metabolism, isolated from the biological metabolism in a closed loop system. Technical nutrients are used in "products of service," durable products for which the manufacturer retains ownership and leases the product services to customers. In this way, customers get the use of the product without the liability for disposal after use. The manufacturer designs the product so that biological and technical nutrients can be separated and used again to make new products. By keeping the technical nutrients separated from the biological nutrients, the technical nutrients can be "upcycled" instead of recycled, that is, they maintain the original quality of the nutrient.[48]

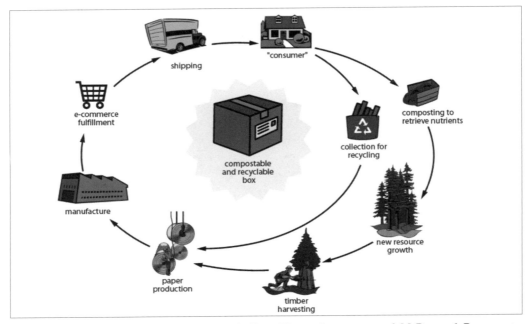

Figure 2-11: Eco-effectiveness: biological metabolism. Illustration courtesy of McDonough Braungart Design Chemistry (MBDC). Illustration copyright 2003, McDonough Braungart Design Chemistry (MBDC). All rights reserved. Do not use without express written permission from MBDC. Contact: 434/295-1111 (v); info@mbdc.com (e).

In their explanations of eco-effectiveness, McDonough and Braungart take eco-efficiency to task, depicting it as an effort to direct an intrinsically destructive system to be less destructive. In their rationale for their eco-effectiveness model, they seem to ignore the fact that sustainability is a journey involving the overhaul of an entire infrastructure.

The WBCSD, the group that invented and developed the eco-efficiency concept, takes issue with McDonough and Braungart's criticism. It notes that, in effect, their proposed solution to the problems of eco-efficiency was to change the name to eco-effectiveness![49] The WBCSD has a point. Read carefully, eco-effectiveness is eco-efficiency taken to its logical endpoint, while sidestepping those annoying practicalities such as a huge legacy manufacturing infrastructure, compulsory standards and specifications, and consumer habits and expectations.

Industrial ecology

Industrial ecology is defined as "*the study of the physical, chemical and biological interactions and interrelationships both within and between industrial and ecological systems* (emphasis added)."[50] Under this concept, industrial systems work not in isolation, but in concert with ecological systems. It is described, perhaps oversimplified, as the science of sustainability, taking an objective look at the relevant issues and practices without judging whether these matters are good or bad.[51]

One of the goals of industrial ecology is to shift the current linear production system to one that emulates biological models, that is, where the wastes from one industrial process are used as feedstocks for another. Industrial systems evolve through a series of stages:

- *Type I system.* This system assumes unlimited resources and unlimited carrying capacity (sinks) for waste disposal. Raw materials and energy resources are used by the ecosystem component and produce either products or by-products (wastes). Since the resources and sinks are not infinite, this system is not sustainable.

- *Type II system.* Analogous to today's industrial system, material flows within the ecosystem are quite large, however, the input and exit streams are somewhat reduced.

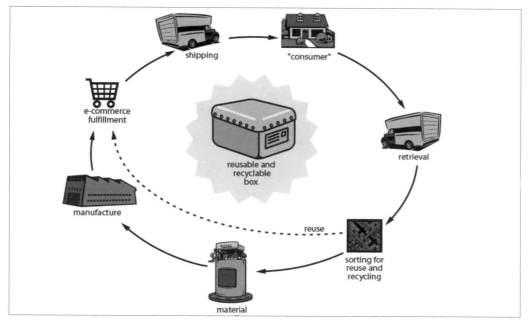

Figure 2-12: **Eco-effectiveness: technical metabolism. Illustration copyright 2003, McDonough Braungart Design Chemistry (MBDC). All rights reserved. Do not use without express written permission from MBDC. Contact: 434/295-1111 (v); info@mbdc.com (e).**

Although significant products, by-products, and waste materials have been reused or recycled, the system is still unsustainable.

- *Type III system.* This is a system in a state of sustainability. This is a totally closed system in which renewable energy is the only input. All products and by-products are reused and recycled.

The industrial park at Kalundborg, Denmark, is famous among industrial ecologists as a demonstration of an integrated system that makes extensive use of industrial by-products and wastes. Here six companies—a waste management organization and the Kalundborg municipality—are using the by-products from one another on an environmentally and financially sustainable basis.

While Kalundborg is a showcase example of how industrial ecology can work in practice, it brings up the question of model resiliency, that is, how does the model cope with marketplace changes that affect energy and material flows in the system? It seems that a robust design, including many system buffers, is an essential ingredient for an ecoindustrial park to be successful.

By-product synergy

By-product synergy emerged in the early 1990s when Chaparral Steel and Texas Industries, Inc. (TXI) found an opportunity to integrate the steel and cement production processes in a way that had not been done before. Similar to the Kalundborg example, Chaparral Steel and TXI found that the slag coming from the steel-making process could be used to make high-quality portland cement. Furthermore, the resulting cement process produced about 10 percent more cement, while using substantially less energy and creating less nitrogen oxide emissions. Last but not least, this arrangement increased profits for both companies.

The success of this cross-industry collaboration led Chaparral to join forces with the Business Council for Sustainable Development, Gulf of Mexico, a regional partner of the World Business Council for Sustainable Development. The result was by-product synergy (BPS), a practice in which companies from different industry sectors share information on their feedstock needs as well as unwanted by-products. What the council, Chaparral Steel, and TXI figured out

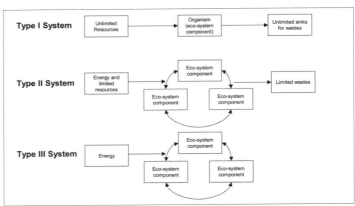

Figure 2-13: Industrial ecology model systems. Adapted from Braden R. Allenby, Industrial Ecology: Policy Framework and Implementation. (Upper Saddle River, NJ: Prentice Hall, 1999), 44.

was that while process engineers are very familiar with the operations within their own company and industry, they are not familiar with operations in other industries. They operate in relative isolation, rarely exchanging information across industry sectors.

After learning about the success of Chaparral Steel and TXI on a single BPS opportunity, the council decided to pilot a larger BPS project. The council brought the BPS concept to its member companies in Tampico, Mexico, and began a yearlong project with 20 companies participating. The participants expected to find two or three synergy opportunities. Instead they identified 65. Out of these, 29 appeared to have immediate commercial potential. Thirteen were ultimately pursued.

The BPS process is a facilitated information exchange among the engineering and operations staff of the participating companies. These companies are drawn from a small geographic region, perhaps within a 50-mile radius. The reason for this choice is not only to

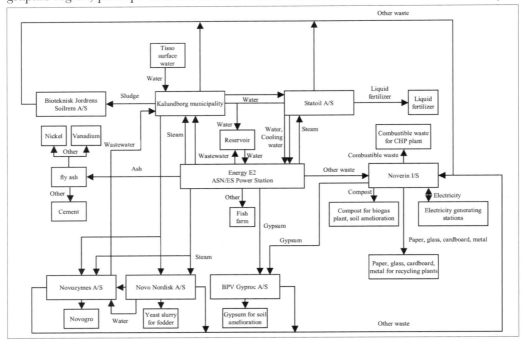

Figure 2-14: Kalundborg eco-industrial park. Adapted from "Industrial Symbiosis: Exchange of Resources," http://www.symbiosis.dk/.

keep travel costs low but to recognize the transportation costs associated with any of the synergies identified. During the meetings, the participants are exposed to the production processes, raw material needs, and the waste streams of the neighboring companies. Through extensive collaboration and a systematic analysis of feedstocks and waste streams, the participants identify potential synergies. These are further refined in order to distill those most commercially promising.

On the surface, BPS resembles the old waste exchange programs created by the EPA in the 1970's. These programs were a kind of industrial garage sale, a published list of available industrial by-products offered in batch quantities to the open market. Since the quality was not specified or consistent, and the quantities were variable, this early effort did not instill much confidence with the users. As a result, the programs had little success.

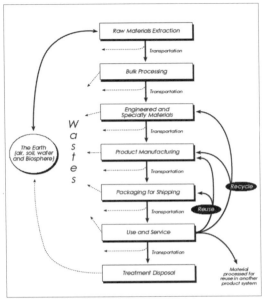

Figure 2-15: Design for Environment (DfE): product life cycle. Source: Jeremy M. Yarwood and Patrick D. Eagan, *Design for the Environment: A Competitive Edge for the Future Toolkit*. Minnesota Office of Environmental Assistance. p. 7.

Unlike these early waste exchange programs, the BPS process recognizes that the acceptance of one company's waste as another company's feedstock takes more than simply announcing its availability. Rather, it is a relationship-building process between individuals and companies. Success is predicated on the establishment of a high degree of trust: that the waste or by-product stream can be delivered reliably and within manageable specifications of quantity and quality. To these ends, the BPS process is both an information collection and a facilitation process, bringing information and people together for the first time from a diverse set of industries to reveal and discuss feedstock needs and the by-products produced. As the BPS process proceeds, participant relationships are formed and barriers begin to fall, opening up new opportunities.

Design for Environment (DfE)

Design for Environment (DfE) is a systematic approach for incorporating environmental considerations into the entire product life cycle, starting with the selection of materials, through manufacture, packaging, consumer use, and disposal. Following Bill McDonough's and Paul Hawken's astute observations that our environmental problems are the result of faulty design, DfE recognizes that many, if not most, of the sources of waste and pollution could be substantially reduced or eliminated at the product design stage. DfE addresses this issue by integrating environmental impact to the existing set of design considerations.

The push for DfE arose out of new regulations and increasing customer expectations that the manufacturers of the products they buy have thought through the potential environmental impacts of those products. As a result, a growing number of manufacturers are applying DfE concepts to their product designs. In addition to reducing overall environmental impact, manufacturers are learning that they can save money by reducing material usage and energy consumption as well as future liabilities.

There are three unique characteristics of DfE:

- The entire life cycle of a product is considered. This includes the resources used to make the product, all the way through to product disposal.

- It is applied early in the product realization process. DfE is brought into the concept stage, and it is considered along with conventional design considerations such as customer requirements, manufacturability, economics, function, and weight and size.

- Decisions are made using a set of values consistent with industrial ecology, integrative systems thinking, or another framework. DfE recognizes the overall issues of sustainability, that the continued use of nonrenewable resources and reductions to carrying capacity are not sustainable in the long term. Accordingly, DfE methodologies incorporate design principles that are consistent with accepted sustainable development principles such as eco-efficiency, eco-effectiveness, the Natural Step, natural capitalism, and others.[52]

There are many examples of DfE successes:

- Xerox has saved literally hundreds of millions of dollars by changing its copier designs so that parts are reusable and easily repaired.

- Hewlett-Packard uses modular architecture to make its printers easier to repair. The company also designed its printers using less materials as well as recycled plastics from old telephones. Its new line of ink-jet printers use 80 percent less power than dot matrix printers.[53]

- Electrolux has redesigned its apartment-scale washing machines, optimizing the use of electricity, water, and detergent. These changes have reduced laundry costs by 50 percent.[54]

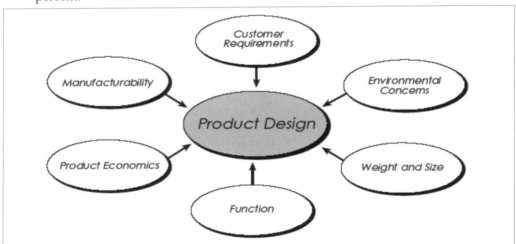

Figure 2-16: Design for Environment (DfE): product design considerations. Source: Design for Environment (DfE): product life cycle. Source: Jeremy M. Yarwood and Patrick D. Eagan, *Design for the Environment: A Competitive Edge for the Future Toolkit*. Minnesota Office of Environmental Assistance. p. 9.

To some engineers, the principles of sustainable development may seem too theoretical and the problems too distant to be useful in their daily work. To that group, I suggest they focus less on the high level principles and more on the opportunities. The next chapter focuses on sustainable development as a market driver, creating new markets and new engineering services.

---

[1] Daniel Sitarz, ed., Agenda 21: The Earth Summit Strategy to Save Our Planet (Boulder, CO: Earthpress, 1994), 4–5.

[2] World Commission on Environment and Development, Our Common Future (New York: Oxford University Press, 1987).

[3] World Economic Outlook (Washington, DC: The International Monetary Fund, May 1999), 22.

[4] World Economic Outlook: Growth and Institutions (Washington, DC: The International Monetary Fund, April 2003), 7.

[5] Richard Weingardt, Forks in the Road, 1.

[6] Ibid., 10.

[7] Paul Hawken, The Ecology of Commerce (New York: HarperBusiness, 1993), xiii.

[8] Daniel Sitarz, ed., Sustainable America (Carbondale, IL: Earthpress, 1998), 24.

[9] Garrett Harden, "The Tragedy of the Commons," Science 162 (1968):1243–1248.Writing in Science, Garrett Harden explains the concept of the tragedy of the commons this way. "Picture a pasture open to all. It is to be expected that each herdsman will try to keep as many cattle as possible on the commons. … Each herdsman seeks to maximize his gain. Explicitly or implicitly, more or less consciously, he asks, 'What is the utility to me of adding one more animal to my herd?' This utility has one negative and one positive component. The positive component is a function of the increment of one animal. Since the herdsman receives all the proceeds from the sale of the additional animal, the positive utility is nearly + 1. The negative component is a function of the additional overgrazing created by one more animal. Since, however, the effects of overgrazing are shared by all the herdsmen, the negative utility for any particular decision--making herdsman is only a fraction of − 1. Adding together the component partial utilities, the rational herdsman concludes that the only sensible course for him to pursue is to add another animal to his herd. And another. … But this is the conclusion reached by each and every rational herdsman sharing a commons. Therein is the tragedy. Each man is locked into a system that compels him to increase his herd without limit—in a world that is limited. Ruin is the destination toward which all men rush, each pursuing his own best interest in a society that believes in the freedom of the commons. Freedom in a commons brings ruin to all."

[10] Credit is graciously given to Don Roberts, who first offered this model in an earlier form in his paper "Sustainable Development—A Challenge for the Engineering Profession" at the International Federation of Consulting Engineers (FIDIC) conference in Oslo, Norway, in 1990.

[11] Herman Daly and John Cobb Jr., Redirecting the Economy Toward Community, the Environment, and a Sustainable Future, (Boston: Beacon Press, 1989).

[12] "Rachel's Teachings, Rachel's Warnings," Center for Health, Environment and Justice, www.chej.org. Previously published in Everyone's Backyard 14, no. 2 (Summer 1996), http://www.chej.org/ORGBOX/rachels.html.

[13] "The Story of Silent Spring," Natural Resources Defense Council, http://www.nrdc.org/health/pesticides/hcarson.asp.

[14] Peter Matthiessen, "Time 100: Scientists and Thinkers—Rachel Carson," http://www.time.com/time/time100/scientist/profile/carson.html.

[15] "What Is Sustainable Development?" Commissioner of the Environment and Sustainable Development, Canada, http://www.oag-bvg.gc.ca/domino/cesd_cedd.nsf/html/menu6_e.html.

[16] "Chapter 2: State of the Environment and Policy Retrospective: 1972–2002," GEO: Global Environmental Outlook 3, http://www.grida.no/geo/geo3/english/081.htm#tab31.

[17] Gro Harlam Brundtland, quoted by Andy Duncan, "This Norwegian's past may connect with your future" http://oregonfuture.oregonstate.edu/part1/pf1_03.html.

[18] World Commission on Environment and Development, Our Common Future, 8.

[19] Brandt 21 Forum, http://brandt21forum.info/summary.htm.

[20] "Science for Sustainable Development," in Agenda 21 (New York, NY United Nations, 1987), Section 35.2.

[21] "Information for Decision-making," in Agenda 21, Section A.

[22] These ecosystems include agroecosystems, forest ecosystems, freshwater ecosystems, grassland ecosystems, and coastal and marine ecosystems.

[23] Eugene Linden, "State of the Planet: Condition Critical," Time, April-May 2000, 19.

[24] "Millennium Ecosystem Assessment Concept Paper," http://www.millenniumassessment.org/en/about/concept.htm.

[25] "Johannesburg Summit 2002: Key Outcomes of the Summit," www.johannesburgsummit.org.

[26] John Elkington, Cannibals with Forks: The Triple Bottom Line of 21st Century Business (Gabriola Island, BC, Canada: New Society Publishers, 1998). The provocative title derives from a question posed by the Polish poet Stanislaw Lec: "Is it progress if a cannibal uses a fork?"

[27] Jan Bebbington and Rob Gray, "Sustainable Development and Accounting: Incentives and Disincentives for the Adoption of Sustainability by Transnational Corporations," Environmental Accounting and Sustainable Development: The Final Report (The Netherlands: Limperg Institute, 1996). Quoted in "Elkington, John", p. 79.

[28] "Elkington, John", Cannibals with Forks, 69–96.

[29] The WBCSD is an international business organization with a core membership of 165 companies, all committed to the development and promotion of the business case for sustainable development. In addition to the core members, it has a regional network of 43 national and regional business councils and partner organizations located in 39 countries. Its Web site is http://www.wbcsd.org.

[30] Björn Stigson, WBCSD Strategy 2010 (Geneva, Switzerland, World Business council fo r Sustainable Development , June 2001).

[31] Brian Nattrass and Mary Altomare, The Natural Step for Business (Gabriola Island, BC, Canada: New Society Publishers, 1999), 23.

[32] Paul R. Ehrlich, "Ecological Economics and the Carrying Capacity of the Earth," in Investing in Natural Capital, ed. Ann Marie Jansson, Monica Hammer, Carl Folke, and Robert Costanza, 43 (Washington, DC: Island Press, 1994).

[33] Brian Nattrass and Mary Altomare, The Natural Step for Business, 18.

[34] Ray C. Anderson, Mid-Course Correction (Atlanta: The Peregrinzilla Press, 1998), 19. Ray credits William Mc-Donough, then dean of the School of Architecture of the University of Virginia, for the concept. Ray Anderson is the chairman and CEO of Interface, Inc., the world's largest producer of commercial floor coverings.

[35] The Natural Step Canada, "The Natural Step Framework for Sustainability," http://www.naturalstep.ca/implementation.html.

[36] Amory B. Lovins, L. Hunter Lovins, Paul Hawken, "A Road Map for Natural Capitalism," 146.

[37] Ibid.

[38] Paul Hawken, Amory Lovins, and L. Hunter Lovins, Natural Capitalism.

[39] Amory B. Lovins, L. Hunter Lovins, Paul Hawken, "A Road Map to Natural Capitalism," 149–150. Speaking to interviewer Kirsten Garrett in 2000 on Australia's ABC Radio National, Lovins noted that despite the savings, the building owner never made the change. It seems that the property was controlled by the leasing agent, who didn't want to delay her commissions while the building was being refurbished.

[40] Janine M. Benyus, Biomimicry: Innovation Inspired by Nature (New York: William Morrow and Company, 1997), 135.

[41] Amory B. Lovins, L. Hunter Lovins, Paul Hawken, "A Road Map to Natural Capitalism," p. 156.

[42] The Business Council for Sustainable Development was the predecessor organization to the World Business Council for Sustainable Development.

[43] Stephan Schmidheiny, Changing Course (Cambridge, MA: MIT Press, 1992).

[44] Livio D. Simone, Frank Popoff, Eco-efficiency, 47–56.

[45] Scott Johnson, "The Economic Case for High Performance Buildings," 353.

[46] Ibid, 356–57.

[47] William McDonough and Michael Braungart, "The NEXT Industrial Revolution," Atlantic Monthly, October 1998: 82–92.

[48] William McDonough and Michael Braungart, Introduction to Cradle to Cradle Design Framework, CD-ROM, version 7.02, McDonough Braungart Design Chemistry, 2002.

[49] Charles O. Holliday, Stephan Schmidheiny, and Philip Watts, Walking the Talk, 86.

[50] Andy Garner and Gregory A. Keoleian, Industrial Ecology: An Introduction (Ann Arbor: National Pollution Prevention Center for Higher Education, University of Michigan, 1995), 2.

[51] Braden R. Allenby, Industrial Ecology: Policy Framework and Implementation (Upper Saddle River, NJ: Prentice Hall, 1999), 40–41.

[52] Jeremy M. Yarwood and Patrick D. Eagan, Design for the Environment: A Competitive Edge for the Future Toolkit (St. Paul, Minnesota Office of Environmental Assistance), 6.

[53] Yarwood and Eagan, Design for the Environment, 12.

[54] Electrolux Environmental Report, 1998, as reported in http://dfe-sce.nrc-cnrc.gc.ca/overview/benefits_e.html.

CHAPTER 3

Sustainable Development:
Client Needs and Market Drivers

Poudre School District is committed to being a responsible steward of our natural resources and believes that public education should provide leadership in developing an ethic of sustainability in all of its practices. We recognize that sustainable design may require a fundamental shift from certain aspects of conventional design and construction. However, we stand committed to sustainable design and are confident it will yield positive outcomes for our students and the community. [1]

"Where's the beef?"

Following that now-famous and penetrating inquiry made by Clara Peller, the former pitch-woman for Wendy's restaurants, many of my colleagues are asking a similar question: *If sustainable development is such a hot idea, why haven't I seen any sustainable development projects?* My response has been this: (1) you may not have recognized the relationship of the project to sustainable development or, worse, (2) you may not have been asked to participate.

Corporations, primarily the large multinationals and those producing substantial environmental impacts, are reacting to a set of trends and market forces and changing the way they operate and compete. These changes are initiated at the highest levels of the corporation. To formulate new strategies and operating policies, the CEO, the board, and senior staff work mostly internally. If they seek advice, it is generally from world-class experts like Paul Hawken, Amory Lovins, L. Hunter Lovins, Brian Nattrass, Mary Altomare, John Elkington, and William "Bill" McDonough. These are highly creative people who can develop a compelling future vision of a sustainable organization, along with the corresponding strategies and policies needed to revamp the company's operations.

At the early stages of the changeover, support from engineering companies is usually not solicited, at least not unless the firm has something substantial to contribute in the way of vision and strategy. Requests for proposals for engineering services, when they are eventually issued, involve services to implement the changeover. These solicitations are parceled out to the organization's engineering consultants and usually involve the more conventional services such as environmental upgrades. Solicitations issued in this form make sustainable development themes hard to see. In addition, sustainable development may show up in the form of selection criteria, either explicitly listed in the proposal request or made implicit by the client's project manager. Firms working closely with their clients and understanding the drivers behind the solicitation have a big advantage.

A good example of how an organization's commitment to sustainability translates into engineering services is CH2M HILL's work for Nike. In the late 1990s, Nike rewrote its environmental and social policies, affirming its commitment to sustainable development practices throughout its supply chain. That commitment translated into a number of programs to reduce Nike's impact on the environment, including the reduction of the use of toxic chemicals in its manufacturing plant, an emphasis on recycling, and the upgrading of its water and wastewater facilities to global operating standards, all in concert with Nike's sustainability strategies and policies. CH2M HILL knew the client well and was given the job of upgrading those facilities. Today the firm continues to support Nike's journey toward sustainability.

Without knowing the origins or the driving forces, this work might be viewed as just another good but routine engineering assignment. Could any good water-wastewater engineering firm do this work? Probably. Does an in-depth knowledge of sustainability principles and practices help win the assignment? You bet!

The purpose of this chapter is to describe the trends and market forces that are creating the market for sustainable development services. What are the trends and forces that are causing clients—corporations and government agencies alike—to be concerned about sustainable development? What are these organizations doing about it? How do their concerns and consequent actions create a market for engineering firms in sustainability services? Corporations, government agencies, states, and municipalities are making these changes not only because of stakeholder pressures, but because they make good business sense.

Overall trends and market drivers

How is the issue of sustainable development emerging as a major force in the market? How is it driving both corporations and government organizations to change their overall strategies and policies and declare themselves a sustainable organization? How might this issue affect your clients? This section presents five key trends which, taken together, are pushing organizations to change their operations and commit substantial resources to becoming recognized as being sustainable.

How issues emerge

As is often the case, trends by themselves do not indicate any specific threat or opportunity. Rather, they are indicators of change that, when aggregated together, can indicate the emergence of an issue. The futurist firm of Coates and Jarratt characterized these trends as warning signals—indicators of change that point to a larger emerging issue. It suggested further that issues emerge in stages, beginning with a few barely decipherable trends and events, eventually forming into a well-defined issue on the public agenda. Coates and Jarratt identified three stages, some of which are broken out into sublevels.[2] These are depicted in the table 3-1.

Table 3-1: How issues emerge

Stage	Description
Early warning	
Early awareness	Vague concern about the condition or change
Informal dialogue	Discussion among special interest groups or fringe publications
Gathering knowledge	Experts begin to think, write, and communicate; issue named
Midwarning	
Issue championed	Issue championed and framed in terms that the general public can understand
Public knowledge	Broad public knowledge through publicity, a defining event
Late warning	
Full recognition	Full-blown issue, national press, legislation

Issues start in the early warning stage with a few trends that provide a hazy picture of a possible change. As more information becomes available, there is additional discussion, starting with special interest groups, eventually emerging as a named thesis. The issue moves into the midwarning stage when it is framed and described in layman's terms. In this stage, there is broad discussion with many groups taking sides. In the final stage, the issue reaches the national press. Depending on the solution required, the issue could become at this stage the subject of national legislation. At each stage, the issue may be reshaped one or more times based on new information or events. It follows that issues are more easily shaped in the early stages of development.[3] The emergence of the sustainable development issue is plotted in figure 3-1.

Five key trends

Public- and private-sector organizations are starting to experience the pressures created by the current economic development model. The movement toward sustainable development appears to be driven by five underlying trends:

- Rapid economic growth
- Corresponding expansion of consumption
- Growing awareness of consequences
- Emergence of powerful new stakeholders
- Emergence of credible industry voices

Rapid economic growth

With the emergence of an integrated world economy, the world is poised to grow and develop rapidly in this century. Since the end of the cold war, trade barriers have been dropping. Fueled by population growth in the developing countries and enabled by developments in information technology and telecommunications (IT), we are moving toward a borderless world where companies have the ability to sell anything, anywhere, at any time. Based on this unparalleled openness, companies can now take advantage of scale economies as well as lower labor and operating costs. Many are shifting their production facilities to more economical locations, a large number of which are located in World 2 and World 3 nations. In turn, this shift has generated new jobs and a corresponding growth in income.

For the developing world, this income growth has expanded disposable income, creating new markets for the goods and services that improve and enrich their lives. Trading connections with the developed world, coupled with TV satellite broadcasting and the Internet, have raised people's goals, expectations, and aspirations for a better quality of life.

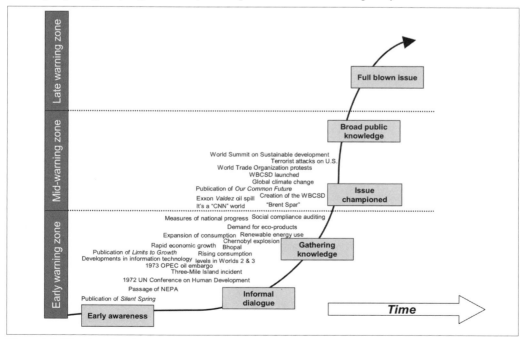

Figure 3-1: Warning signals of the emergence of sustainable developm ent. Adapted from Coates and Jarratt, Sustainability in the 21st Century, Implications for Business 2001–2025, Business Case for Sustainability Briefing (unpublished report, Washington, DC, Spring 2002).

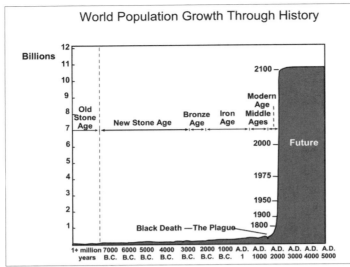

Figure 3-2: World population growth through history.
Source: Population Reference Bureau and United Nations, World
Population Projections to 2100, (United Nations, New York, NY, 1998).

Population expansion into urban areas presents an additional challenge. While the number of megacities[4] in World 1 is expected to increase only to five, the number of megacities in Worlds 2 and 3 is expected to reach over 50. Furthermore, most of these cities will be located in coastal areas, making them highly vulnerable to natural disasters and the other consequences of global climate change.[5] This will create a major challenge for the engineering community, requiring the design and construction of a new kind of infrastructure to support these huge urban populations.

Corresponding expansion of consumption

World population just reached 6 billion and is expected to reach 9–10 billion by the middle of the century. Over the next 25 years, the world's population is expected to increase by 2 billion people. Almost all (99 percent) of that growth will occur in Worlds 2 and 3 and almost all in urban areas.[6] In addition, income growth is expected to grow at an annual rate of 3 percent, translating into a fourfold increase.[7] Today 81 percent of the world population lives in World 2 and 3 countries. This number is expected to increase to 86 percent by 2050.[8]

Here Paul and Anne Ehrlich's environmental impact equation comes into play.[9] First, unprecedented population growth combined with new goals of quality of life to accelerate the demand for goods and services. With a higher disposable income, it is not surprising that people in the developing world are seeking the same conveniences and enjoyments in the form of products and services found in the World 1 counties. Moving toward what they perceive as a better quality of life, they are seeking more mobility (in the form of automobile ownership and increased air travel) and more protein in their diets.[10]

Second, with no other alternatives in sight, they are destined to follow the historical development pathways, that is, using essentially the same take-make-waste technologies that World 1 countries used. These T_1 technologies, as Ray Anderson described them,[11] will have substantial negative impact on resource consumption, pollution generation, and land use.

Growing awareness of the consequences

However, with these improvements come new risks. There is now hard evidence that we are approaching the limits of our resources and ecological carrying capacity at all scales.[12] While improvements in information technology have enabled economic expansion, they have also improved our ability to detect, assess, and communicate resource and ecological problems. Today we can do broad scale ecological assessments using satellite imaging and analysis. Advances in sensor technology and analytical chemistry enable us to detect contamination at a low cost and at extremely low levels. Furthermore, we now have more credible methods of

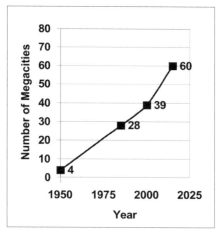

Figure 3-3: Projected growth of mega-cities. Adapted from "Mega Cities and Mega Slums in the 21st Century," WaterAid www.wateraid.org.uk.

assessing risk to human health and the environment. Advances in information technology have provided low-cost communication and information dissemination, not only facilitating collaboration among the scientific and engineering disciplines, but also enabling nongovernmental organizations (NGOs) and public interest groups to access and share this information easily.

Although these threats are clearly ominous signs of a degrading environment, they are generally unheeded. At numerous times in the past, many organizations have issued strong and dire warnings about impending resource shortages and corresponding disasters, only to find that their doomsday scenarios never materialized. Throughout history, new technological developments and transfers have forestalled serious shortages by either increasing productivity dramatically or creating viable substitutes. For example, in what is labeled the *Green Revolution*, Norman Borlaug was able to stave off predictions of massive starvation in Pakistan and India by the introduction of Western-style high-yield agriculture.[13]

However, in the face of unprecedented growth in both population and resource demand, it is questionable whether technology will be able to keep up. In addition, the impacts of some of these new technologies (e.g., the introduction of genetically modified foods, the use of bovine somatotropin [bST] to enhance milk production) are unknown and are raising considerable public concern. These risks were not considered serious, at least until now.

Emergence of new and powerful stakeholders

With these advances in communication and information dissemination, new and powerful stakeholders have emerged. In this new "CNN world," NGOs and public interest groups have unprecedented access to corporate and government performance, good and bad. Moreover, they possess new information technology tools for taking action. They can deliver their information and perceptions about corporate performance to a worldwide audience instantaneously and at very low cost. These actions can affect a company's reputation and hence its bottom line. In effect, these groups are setting the standards for corporate economic, environmental, and social performance.

As never before, industries and government agencies alike are feeling pressures from stakeholder groups. These groups come armed with new information, previously unobtainable, about the performance of the organization, cast in terms of what their constituencies need and value. They also come equipped with new communication tools—easy, low-cost electronic communications to a network of consumers, legislators, the media, and other like-minded groups literally around the world. These groups can easily track corporate performance and rate it against their own organization's norms. They can also communicate those results and suggest an appropriate response to anyone on the electronic network at essentially zero cost. Just how powerful is this electronic network? Think of it this way. According to Moore's law, computing power, rated as performance per unit cost, doubles every 18 to 24 months.[14] Furthermore, increases of this magnitude have been going on for several decades, and there is no end in sight. For those of us who use computers, we have grown to expect this increase, often much to our chagrin, since it means a continuing expense of computer upgrading, also with no end in

Table 3-2: The top 30 megacities

2000		2015 (estimate)	
City	Population, millions	City	Population, millions
1. Tokyo, Japan	34,450	1. Tokyo, Japan	36,214
2. Mexico City, Mexico	18,066	2. Mumbai (Bombay), India	22,645
3. New York, United States	17,846	3. Delhi, India	20,946
4. São Paulo, Brazil	17,099	4. Mexico City, Mexico	20,647
5. Mumbai (Bombay), India	16,086	5. São Paulo, Brazil	19,963
6. Calcutta, India	13,058	6. New York, United States	19,717
7. Shanghai, China	12,887	7. Dhaka, Bangladesh	17,907
8. Buenos Aires, Argentina	12,583	8. Jakarta, Indonesia	17,498
9. Delhi, India	12,441	9. Lagos, Nigeria	17,036
10. Los Angeles, United States	11,814	10. Calcutta, India	16,798
11. Ōsaka-Kōbe, Japan	11,165	11. Karachi, Pakistan	16,155
12. Jakarta, Indonesia	11,018	12. Buenos Aires, Argentina	14,563
13. Beijing, China	10,839	13. Cairo, Egypt	13,123
14. Rio de Janeiro, Brazil	10,803	14. Los Angeles, United States	12,904
15. Cairo, Egypt	10,398	15. Shanghai, China	12,666
16. Dhaka, Bangladesh	10,159	16. Metro Manila, Philippines	12,637
17. Moscow, Russian Federation	10,103	17. Rio de Janeiro, Brazil	12,364
18. Karachi, Pakistan	10,032	18. Ōsaka-Kōbe, Japan	11,359
19. Metro Manila, Philippines	9,950	19. Istanbul, Turkey	11,302
20. Seoul, Republic of Korea	9,917	20. Beijing, China	11,060
21. Paris, France	9,693	21. Moscow, Russian Federation	10,934
22. Tianjin, China	9,156	22. Paris, France	10,008
23. Istanbul, Turkey	8,744	23. Tainjin, China	9,874
24. Lagos, Nigeria	8,665	24. Chicago, United States	9,411
25. Chicago, United States	8,333	25. Lima, Peru	9,365
26. London, United Kingdom	7,628	26. Seoul, Republic of Korea	9,215
27. Lima, Peru	7,454	27. Santa Fe de Bogotá, Colombia	8,900
28. Tehran, Iran	6,979	28. Lahore, Pakistan	8,699
29. Hong Kong, China	6,807	29. Kinshasa, Democratic Republic of the Congo	8,686
30. Santa Fe de Bogotá, Colombia	6,771	30. Tehran, Iran	8,457

Source: World Urbanization Prospects: The 2003 Revision, Data Tables and Highlights ESA/P/WP.190 (New York, NY, United Nations, Department of Economic and Social Affairs, Population Division, March 24, 2004).

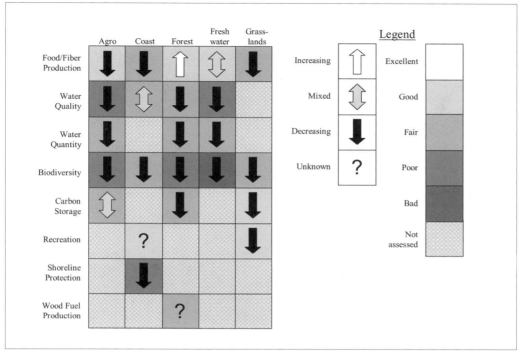

	Agro	Coast	Forest	Fresh water	Grass-lands
Food/Fiber Production	↓	↓	↑	↕	↓
Water Quality	↓	↕	↓	↓	
Water Quantity	↓		↓	↓	
Biodiversity	↓	↓	↓	↓	↓
Carbon Storage	↕		↓		↓
Recreation		?			↓
Shoreline Protection		↓			
Wood Fuel Production			?		

Legend — Increasing ↑ Excellent; Mixed ↕ Good; Decreasing ↓ Fair; Unknown ? Poor; Bad; Not assessed

Figure 3-4: PAGE ecosystem assessment. Source: World Resources 2000-2001: People and Ecosystems: The fraying web of life. United Nations Environment Programme, World Resources Institute, Washington, DC, 2001. p. 47.

sight. But this power increase only considers individual personal computers. When you hook personal computers to a network, power and corresponding value increase dramatically. Bob Metcalfe is credited with deriving what today is known as Metcalfe's law of networks. Working on data communications in the Hawaiian Islands, Metcalfe found that the potential value of a network, whether it is telephones, faxes, or computers, increases by the square of the number of connections. That is, connect n number of computers together and you have a potential value of n times n. When you consider the tens of millions of computers connected to the Internet, tens of millions squared is a huge number.[15]

Kevin Kelly, founding editor of *Wired* magazine, thinks Bob Metcalfe may be too conservative. Writing in his book *New Rules for the New Economy*,[16] Kelly notes that Metcalfe's law really applies to point-to-point network connections, like telephones and fax machines. On the Internet, where people can make multiple, simultaneous n connections between groups of people, the value is not n times n. Rather, it is more like n^n, or n to the nth power.

Given all that potential communication power at little or no cost, stakeholders have tremendous leverage. People who do not even know each other can use the Internet to organize around a common theme of poor corporate performance. Then they can take actions to alter public opinion or initiate a marketplace response. Author Howard Rheingold calls them "smart mobs," people who are able to act in concert, even though they do not know each other.[17] Organizing through the Internet and mobile telephones, such groups have successfully

Figure 3-5: The power of computer networks. Source: ClipArt.com com and JupiterImages.

protested and disrupted international meetings of major organization such as the World Trade Organization.

Emergence of credible industry voices

The old argument that "good environmental and social performance is too expensive" is no longer convincing. Successful companies like Dow Chemical Company, DuPont, BP, and Shell are making a substantial commitment to sustainable development. These companies are members of the World Business Council for Sustainable Development (WBCSD), a member-led organization of over 160 world-class companies, all publicly committed to sustainable development. The sum of the WBCSD members' annual gross revenues, if correlated to gross domestic product, would make it the third-largest economy in the world, close behind the United States and Japan.

At the September 2003 meeting of the WBCSD, over 150 liaison delegates and representatives from 107 of the member companies were asked to give their views on their company's business case for making a commitment to sustainable development. Three-quarters of the participants stated that their sustainable development activities would lead to better stock performance, if not immediately, then in five years. Over two-thirds said that their companies' managers were working systematically to make sustainable development part of their operating policies and strategies. Why incorporate sustainability into your corporate or competitive strategy? The survey participants said that "risk reduction" and "market opportunities" were the primary reasons. The detailed responses to this question are shown in table 3-3.[18]

Supporting evidence from the leading business schools

The relationship between sustainability and business success has been the subject of study at many of the nation's top business schools. The findings track with the views of the WBCSD membership: not only does it pay to be green, but it can be the source of competitive advantage.

Harvard Business School professor Michael E. Porter, one of the world's top experts on competitive strategy, has studied the relationships between sustainability and the competitiveness of nations. Working with Daniel Esty of Yale University, they showed a direct correlation between national competitiveness and environmental sustainability. The authors took issue with the conventional wisdom that countries with weak systems of environmental stewardship fare better economically because they can attract industrial development through lower environmental compliance costs. What they found was just the opposite. National competitiveness is strongly correlated with environmental sustainability, measured as a combination of the state of the country's environmental systems, the level of environmental stress, human vulnerability to environmental impacts, social and institutional capacity to respond to pollution and natural resource problems, and global stewardship. [19]

Table 3-3: WBCSD members' views on the business case for sustainable development

Reason	Percentage
Risk reduction	31%
Market opportunities	19%
Operational efficiency and effectiveness	18%
Enhancement of brand and creation of goodwill	13%
The meaning of life	8%
Protecting the resource base of raw materials	6%
Recruitment and retention of talent	6%

Writing in the July-August 1999 issue of the *Harvard Business Review*, Professor Forest Reinhardt presents five approaches for creating shareholder value through environmental management:

- *Differentiating products.* Create products that deliver credible environmental benefits or lower environmental costs.

- *Managing competitors.* Change the rules of the game so that they place competitors at a disadvantage. Set environmental or corporate social performance standards that make sense and that competitors will have difficulty matching.

- *Saving costs.* Make internal cost reductions while at the same time improving environmental performance. Invent waste-cutting measures and recycling and reuse programs that save money and lower product costs.

- *Managing environmental risk.* Find ways to avoid the costs of stakeholder opposition, pollution, chemical spills, and industrial accidents through education, performance incentives, and full-cost accounting.

- *Redefining markets.* Redefine the business model. Following the principles of eco-efficiency, figure out what customers really want. Do not sell the product; sell the service that the product offers.[20]

Business schools also see sustainability as a source of corporate growth through innovation. In a recent article in the *MIT Sloan Management Review*, authors Stuart Hart and Clayton Christensen point out the extraordinary growth and profit opportunities that exist in filling the unmet needs of the 4 billion people in the World 2 nations. Innovative companies like Grameen Telecom (mobile telephones), Galanz (small, low-cost microwave ovens), and Rolltronics (low-power semiconductor circuits) are creating new products for this huge market that makes up the lower part of the economic pyramid.[21]

Sustainability as a leading indicator of good financial performance

Sustainability is becoming increasingly viewed by the financial community as a key element of long-term investment strategies. Investors are recognizing that the components of sustainability—economic, environmental, and social—combine to influence customer demand, legislation, innovation, stakeholder relations, corporate reputation, and other key parameters for investment decisions. In addition, new trends and issues that influence these parameters seem to appear with increasing frequency. In just the past few years, issues such as global climate change, labor conditions, and corporate transparency have made their way to the forefront of importance.[22]

In response, investment organizations have started high-quality funds investing in companies that meet strict criteria for social and environmental responsibility. In 1999, Dow Jones and Company established the Dow Jones Sustainability Index, a measure of the ability of a company to achieve long-term shareholder value by capturing the market potential for sustainability products and services while reducing costs and risks. This stock index is currently performing above the mainstream market average. From September 2002 to September 2003, the value of the Dow Jones Sustainability Index's World index increased by 23.1 percent while the Dow Jones Global Index, World went up by 22.7 percent.[23]

Figure 3-6: Photo of a ram pump used to carry drinking and irrigation water from a river to a small village in Belize. The pump is easy to operate and maintain. Photo courtesy of Engineers Without Borders–USA.

Market response

These trends and their resulting impacts are driving public- and private-sector organizations toward sustainability. Not only are they changing the way organizations operate, but they can improve competitiveness and create cost savings. They also open up white space opportunities that extend from an organization's core competencies. Market drivers and organizational responses appear to be in six categories:

- Prerequisite for market entry

- Market diversification

- Differentiation

- Process improvement

- Ethical imperative

- Cost savings

Prerequisite for market entry

Large multinational corporations, particularly those in resource-intensive industries or industries that have large environmental footprints, are finding that sustainability is a prerequisite for growth and market entry. Companies seeking to expand into developing countries are finding that the host country is expecting them to apply World 1 standards of environmental performance for operating in their country. They are also expected to deliver socially related services that are usually the responsibility of governments. Success in these markets depends to a large extent on a company's reputation and consistent environmental and social performance, measured against emerging global sustainability reporting standards. In response, these companies are extending the scope and breadth of their environmental management systems and upgrading their facilities. They are also pushing sustainability up the supply chain by requiring their suppliers to take similar actions.

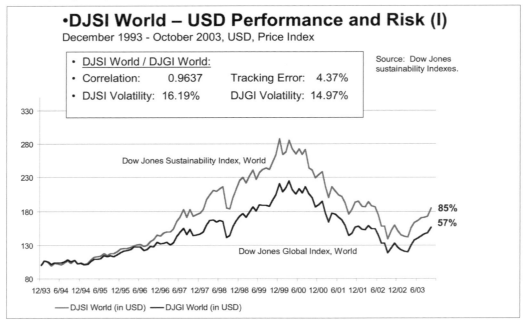

Figure 3-7: Dow Jones Sustainability Index (DSJI): comparison with general world market performance. Illustration courtesy of Dow Jones Sustainability Indexes.

So too with products. Faced with concerns over product success in the market, DuPont and other WBCSD member companies supported a multiyear working group study on innovation, technology, sustainability, and society. The premise for this working group was that innovation, particularly technological innovation, is and always has been a major contributor to the advancement of our quality of life. However, society today is playing an increasingly active role in determining whether or not a particular technological development will be acceptable. In turn, this judgment of acceptability determines the product's success and hence the return on the developer's investment. The study concluded that business needed to better anticipate and understand societal concerns and to find more effective ways to incorporate those concerns into their innovation processes.

Market diversification

The movement toward sustainable development is giving rise to entirely new markets in unexpected regions of the world. Companies such as Procter and Gamble and Unilever are finding new markets in the developing world, creating products and services that meet the special needs of the world's other 5 billion people. The Dow Jones Sustainability Index recognizes Procter and Gamble as the sustainability leader in the household products sector. The company is developing products that deliver safe drinking water and better child nutrition to the World 2 and World 3 nations.

Advances in information technology combined with new business models have enabled companies to produce, package, price, and distribute products and services that meet the special characteristics and pricing requirements of these new customers—low disposable income, limited living space, and less-than-adequate public infrastructure.

Differentiation

For companies such as Stonyfield Farms and Interface Corporation, a commitment to environmental and social good is a major component of their marketing strategy. In addition, cities and counties compete for jobs and growth on the basis of quality of life, and they are using sustainable development concepts to define their current situation and to track progress. As noted earlier in this book, over 20 cities in the United States are seriously pursuing sustainability policies. Seattle, Washington, Austin, Texas, and others have developed complete sets of sustainability measures based on local conditions and stakeholder

Figure 3-8: Industry representatives meet to discuss their By-Product Synergy project in North Texas. Photo by Bill Wallace, Wallace Futures Group, LLC.

values. Marketing on the basis of an organization's sustainability performance requires regular and credible demonstration of economic, environmental, and social performance based on stakeholder values and valid comparisons against the competition.

Process improvement

Sustainability provides a different lens through which companies can view their operations. Several years ago, I asked an executive at 3M about how it developed a business case for its sustainable development policies and practices. He told me that the vision of 3M is to be the most innovative company in its industry. Its goal each year is to have a substantial portion of its revenues generated by new products. It sees sustainability as a catalyst for innovation, enabling it to look at markets and customer needs through a different set of lenses. The result has been a stream of products that use less materials and toxic substances, incorporate recycling and pollution prevention, and more.

By-product synergy is another example of how sustainability can drive industrial process innovation. As described in chapter 2, by-product synergy is a facilitated process in which companies from a number of industry sectors get together and exchange information about unwanted by-products and feedstock needs. In this transitional form of industrial ecology, companies are finding new uses for materials once slated for costly disposal. Here, a new kind of process engineering is required, one that reaches across industry sectors to secure viable feedstock materials in the needed quantity and quality.

Ethical imperative

A number of companies have embraced the principles of sustainable development because it is the "right thing to do." Most notable is Interface Corporation, whose CEO Ray Anderson is striving to transform it into a totally sustainable enterprise. Inspired by Paul Hawken, Daniel Quinn, and others, Interface is seeking to turn the industry model around 180 degrees. Instead of selling material (carpet), it intends to lease floor-coverings. To implement this vision, Interface has established an "Eco Dream Team" composed of the leading thinkers on sustainability. The goal of Interface is to show, by 2020, what it means to be a sustainable corporation.

Cost savings

One of the concerns voiced about the application of sustainability to a company's or government agency's operations is that it adds extra expense with no discernable return. But that all depends on how you do the measurements. Cities like Seattle, Washington, have recognized that the construction cost of a building is only 2 percent of the total cost of operation. Six percent is in operation and maintenance costs, and 92 percent is in the salaries of the people who work in the building. By measuring life cycle costs and benefits, they found that they could realize substantial savings, not only in energy costs, but also in employee productivity and reduced turnover. The Naval Facilities Engineering Command has instituted a program in which it will share with the facility designer the savings generated from reduced facility operating costs.

It is also U.S. Air Force policy to apply sustainable development concepts to all aspects of its facilities and infrastructure projects. In a 2001 policy memorandum issued to all major commands and the organization, the Air Force civil engineer, Major General Earnest O. Robbins II, stated that the Air Force would set goals for its planning, programming, and budgeting process. General Robbins also told the procuring agencies to select engineering consultants based on their qualifications in the various aspects of sustainable development. The memorandum noted that the Leadership in Energy and Environmental Design (LEED) Green Building Rating System was the Air Force's preferred self-assessment metric. It further stated that at least 20 percent of all military construction (MILCON) projects in fiscal year 2004 should be selected as LEED pilot projects, with the goal to have all Air Force MILCON projects be LEED projects by fiscal year 2009. "Sustainable development concepts will benefit the Air Force by creating high performance buildings with long term value," the memorandum stated.[24] In addition, the U.S. Army Corps of Engineers has developed what it calls a Sustainable Project Rating Tool (SPiRiT). This tool parallels the LEED tool, although it has been modified to align with the needs of military installations. According to army policy, all its designs for military facilities shall phase in sustainable design and strive to achieve a SPiRiT bronze level of performance.[25]

Actions by competitors

Are these trends and impacts being recognized? My answer is yes, they are. Unfortunately they are not being recognized all that well by the engineering community. As a result, other competitor organizations are filling the client needs.

Organizations like the Rocky Mountain Institute have established entirely new engineering-oriented practices that assess and remedy engineering designs that were based on conventional metrics. Architectural firms like HOK have established sustainable design groups and have written manuals on sustainable design. Accounting firms like Deloitte Touche Tohmatsu have extended their financial accounting practices into environmental and social reporting in order to meet a growing demand for sustainable development reports. Sustainable development reporting is the mechanism by which companies make public their current status and their intended efforts and investments for meeting sustainability goals. KPMG is engaged in verifying and validating a global carbon dioxide inventory protocol. Organizations such as the Battelle Memorial Institute are being selected to conduct industry studies on sustainability for the mining, cement, and transportation sectors.

A few engineering companies are recognizing the business opportunities in sustainable development and doing something about it. CH2M HILL has established a practice in sustainable development and has a number of projects across the world that incorporate sustainability as a key project goal. These projects have a wide range in scope: planning, facility design, engineering, and training. However, all of these projects have a common theme: the client incorporated sustainable development concepts as part of the scope of the work or the criteria for selection.

The Arup Group Ltd., an international firm based in the Netherlands, has an established sustainability practice that extends worldwide. It provides services in sustainability performance assessment including SPeAR, a four-quadrant model that offers a visual assessment of sustainability. It has been used for evaluating urban redevelopment, manufacturing processes and products, and strategy formulation.[26]

Hatch Engineering, Ltd. of Canada has a substantial sustainable development practice in which it provides a variety of services. It has conducted by-product synergy projects wherein it licenses and services the CemStar process for Texas Industries customers in South America, South Africa, and New Zealand.

Client needs and corresponding opportunities for engineering firms

What are the current opportunities for the engineering community in sustainable development? To start with, there are a number of services that engineering firms can deliver to clients that generally fit within their current core competencies. Summaries of the markets and opportunities associated with sustainable development are presented in table 3-4 at the end of this chapter.

- *Sustainability assessment.* Engineers can assess their client's operation in terms of sustainability performance. They can identify gaps and assist the client in developing a new corporate vision, mission, and goals that incorporate sustainability. Then they can help the client develop strategies and plans to fix or upgrade existing operations. The major competitors are the management consulting firms and the recognized icon consultants in sustainable development. Engineering firms should concentrate on their existing client base, working with clients with whom they have strong relationships with the leadership of the organization.

- *Sustainable development reporting.* Engineers can assist companies in preparing and publishing reports on their status and progress in moving toward sustainability. They can work with a company to determine the needs and interests of its stakeholders regarding company performance. Then they can determine what information to collect and make an assessment against established benchmarks. They can also assist in preparing the report. Finally, they can assist in determining what performance improvements the company should make, design and implement those improvements, and develop the environmental management systems to track performance. The single, most significant competitors for this work are the major accounting firms, who will attempt to parlay their accounting experience (i.e., measuring the company's economic performance) to move into the arena of corporate environmental and social performance measurement. Accounting firms have been successful in capturing significant market share in this area. As the company's auditors, they have developed relationships with senior management. And, at least until recently, they have enjoyed a reputation as a trusted and impartial verifier of corporate performance.

- *Greenhouse gas emissions inventory and reduction.* Engineering firms can assist clients in measuring and assessing their greenhouse gas emissions inventory. This work involves a comprehensive evaluation of existing processes and the calculation of the amount of carbon dioxide or other emissions associated with global climate change. Firms can design and construct new processes and systems for greenhouse gas reduction. They can also help the clients develop greenhouse gas emission offset projects or become involved in emissions trading.

- *Green building design and construction.* This work includes the design and construction of buildings and facilities that incorporate sustainability principles, that is, energy efficiency in lighting, heating, and cooling; reduced water use; and so on. Engineers may also work with clients to achieve a certain level of LEED certification. The major competitors are architectural firms; however, these firms generally do not have the systems integration skills needed to optimize system performance.

- *Industrial ecology and by-product synergy.* Engineering firms can participate in the development and engineering of ecoindustrial parks. These are clusters of industrial facilities specifically designed to match feedstock needs and unwanted by-products. On a broader scale, engineers can assist in organizing and recruiting companies to participate in by-product synergy projects, described earlier in this book. Tasks will include data collection and assessment, selection of promising synergies, feasibility studies, design, and implementation.

Pushing your organization's envelope

In addition to the near-term opportunities listed above, there are several areas worthy of investigation for longer-term opportunities.

Leveraging your knowledge of your clients

Engineering firms have a unique advantage. Because they work for a large number of clients, these firms have been able to amass a substantial client knowledge base spanning a variety of government and industrial sectors. Under the organizing theme of sustainable development, engineering firms should be able to look across industrial sectors, identifying new client needs and inventing new solutions to client problems. In addition, firms should be able to draw upon that breadth of cross-sector experience to advise clients on what others are doing on similar problems.

For example, if copier manufacturers and chemical companies are practicing the principles of eco-efficiency and increasing the service intensity of the products they sell, how can that approach be applied to another client sector? If many of the large multinational companies are producing sustainable development reports, how might that be used to obtain a competitive advantage? Would increased transparency through sustainable development reporting be of any advantage to your client? How might your client use that information to improve its own image or its operations?

Thinking (and working) outside the box

Although image and reputation are still the primary market drivers for corporations and government agencies joining and adopting policies for sustainability, innovation is why they stay and participate. The new paradigm of sustainable development has pushed many companies to question not only how they are conducting their existing operations, but also how sustainability concerns are forcing organizations to think more broadly about new business opportunities.

The question for engineering firms should be this: how might you conduct your engineering practice under the new paradigm of sustainability? Here are some sample questions:

- What new services could you offer if you looked at engineering planning and design problems in terms of life cycle assessment?

- How would you plan and conduct a project if you had good knowledge of stakeholder concerns and issues?

- How might you apply the principles of eco-efficiency or eco-effectiveness?

- Knowing the power of stakeholders in matters of company success, how would you plan and manage a controversial project?

- What approaches for creating shareholder value through environmental management might be applicable to your clients, and how might you present them?[27]

- Under sustainable development, what technologies would be the best to apply and how would you apply them? How would you integrate them with the other technologies and systems involved?

Projects for the bottom part of the pyramid

Companies like Shell, Unilever, and Procter and Gamble see opportunities in marketing their products to World 2 and World 3 nations. But to do so requires a new business model, one that takes into account household income and family living and working conditions. People tend to buy things in small quantities to match the monies available. Their living space is smaller than in World 1 countries so appliances, furniture, and other living necessities are smaller in size. Storage is at a premium.

As noted in the previously referenced article by Hart and Christensen, the ability to sell products and services to the 5 billion people who live in the developing world opens up an entirely new set of markets. However, it means developing an acute awareness of their needs and finding ways to meet them through a manufacturing and distribution system that takes care of the differences.

Table 3-4: Description of markets and opportunities

Markets*	Description	Trends, market forces	Response	Opportunities	Examples
General building	Commercial buildings, offices, stores, educational facilities, government buildings, hospitals, medical facilities, hotels, apartments, housing	• Recognition of available energy and cost savings, increased productivity • Urban sprawl, smart growth • Client desire for green image	• Life cycle costing • Sustainable communities • General Services Administration, Department of Defense adoption of sustainable development principles, practices	• Facilities designed based on sustainability principles • LEED certification for buildings	Ford River Rouge plant, Seattle LEED, U.S. Army Corps of Engineers SPIRIT building policies
Manufacturing	Auto, electronic assembly, textile plants	• Multinational corporations outsourcing to World 2, 3 countries • Economic growth in developing nations • Product take-back legislation (European Union) • Mistrust of corporations • Globalization	• Increased corporate transparency • Proof of sustainable performance • Push toward eco-efficiency • Design for Environment (DfE) • Selling to other 5 billion people	• Environmental management systems incorporating sustainability • By-product synergy • Sustainability reporting • Facility upgrades • Participate in DfE	Nike wastewater treatment, DuPont products for sustainable agriculture
Power	Thermal and hydroelectric power plants, waste-to-energy plants, transmission lines, substations, cogeneration plants	• Technological advances in wind and solar power, fuel cells, hydrogen • United States' drive for energy independence • Concerns over homeland security	• Growth of renewable energy • Hydrogen initiative • Greenhouse gas (GHG) reductions • Protection of power stations, transmission lines	• Wind and solar power • Hydrogen power, fuel cells • GHG assessments, reductions, offset projects	BP, Shell solar and wind power initiatives
Water supply	Dams, reservoirs, transmission pipelines, distribution mains, irrigation canals, desalination and potability treatment plants, pumping stations	• Severe drought in the western United States • Population increase, demographic shifts • Competition for scarce resources • Loss of recreation, fish habitats • Concerns over homeland security	• Building of new water systems • Rehabilitation of old water systems • Deconstruction of dams • Protection of water systems • Development of new water supplies	• Building and rehabilitation of water systems • Water conservation measures • Development of water resources • New designs: rainwater collection, gray water	Water resource development in the western United States
Sewerage and solid waste	Sanitary and storm sewers, treatment plants, pumping plants, incinerators, industrial waste facilities	• Water shortages, drought • Difficulty in siting new landfills • Product take-back legislation (European Union)	• Use of LEED standards, certification • Design facilities based nonsustainable development principles • Increased recycling	• Engineered wetlands • Wastewater recovery systems • DfE: products, packaging	Solid waste reduction measures: 3M, Dell, Nike
Industrial process	Pulp and paper mills, steel mills, nonferrous metal refineries, pharmaceutical plants, chemical plants, food and other processing plants	• Green sourcing • Mistrust in corporations • Concerns over global climate change • Global Reporting Initiative (GRI) reporting • Concerns over homeland security	• Increased corporate transparency • Proof of sustainable performance • Push toward eco-efficiency • GHG reductions • Greening the value chain • Increased security measures	• By-product synergy • GHG assessments, reductions • Sustainability reporting • Facility upgrades • Improvement of eco-efficiency • Security upgrades	BASF eco-efficiency model, TXI's CemStar, Nike Reuse-A-Shoe
Petroleum	Refineries, petrochemical plants, offshore facilities, pipelines	• Middle East unrest • Concerns over global climate change • Growth of alternative energy systems • GRI reporting • Concerns over homeland security	• Investment in renewable energy • Increased corporate transparency • Proof of sustainable performance • Push toward eco-efficiency • GHG reductions • Corporate transparency • Increased security measures	• By-Product Synergy • Sustainability reporting • Facility upgrades • GHG assessments, reductions, offset projects • Improvement of eco-efficiency • Security upgrades	BP, Shell, U.S. Air Force, Public Service Company of Colorado
Transportation	Airports, bridges, roads, canals, locks, dredging facilities, marine facilities, piers, railroads, tunnels	• Concerns over urban sprawl • Increasing smart growth, conservation legislation • Concerns over homeland security	• Shift to multimodal transportation • Renaissance in urban centers • Increased security measures • Context-sensitive design	• Multimodal transportation • Security upgrades	Sustainable cities, antisprawl initiatives, requirements for context-sensitive design
Hazardous waste	Chemical and nuclear waste treatment, asbestos, lead abatement	• Concerns over urban sprawl • Reinventing as a brownfield issue • Environmental justice	• Redevelopment of urban areas • Remediation • Ecosystem recovery	• Brownfield projects • Ecosystem valuation	Numerous brownfield redevelopment sites
Telecommunications	Transmission lines and cabling, towers, antennae, data centers	• Shift to telecom in business, e.g., teleworking, teleconferencing • Developing nations leapfrogging with mobile	• Expansion of telecommunications networks	• Design and build telecom facilities and infrastructure	All telecom companies

As engineers, how might we sell our services to these 5 billion people? Most of the projects of interest to the engineering community have been the large infrastructure projects—roads, bridges, airports, water supplies, wastewater treatment plants, electric power stations—all geared toward delivering the transportation, power, water, and sanitation necessary for the nation's economy. But is this the right model? Can we offer a different model in addition to the former—one that delivers the same services to needy communities that cannot afford the large infrastructure projects and cannot wait for the distribution systems to reach them? In this model, engineers can design new systems that meet the needs of small communities using appropriate technologies, technologies that are simple to install and operate, inexpensive to buy, and easy to maintain.

What does an engineering firm need to do to enter this market? The self-assessment, the decisions, and the changeover that firms need to make if they decide to enter the sustainable development market are discussed in the next chapter.

---

[1] Don Unger, former superintendent of schools, Poudre School District, Fort Collins, Colorado. Presentation to the state of Wyoming, February. The Poudre School District in Fort Collins, Colorado, has won many national awards for its sustainable school designs and operations.

[2] Chris Carbone, "Sustainability in the 21st Century, Report #7: Early Warning Signals" (unpublished report, Coates and Jarratt, Washington, DC, spring 2002).

[3] Ibid., 2–3.

[4] Megacities have various definitions but are generally defined as large cities with a population of 5 million or more, usually with a high population density of at least 2,000 people per kilometer squared. Cities of that size can be seen as advantageous in that the concentration of people is a source of diversity and creativity. However, their population size and geographic breadth create problems in employment, housing, transportation, social services, food, water supply, sanitation, healthcare, security, and environmental quality. Megacities, especially those in the developing world, are vulnerable to natural disasters and other problems associated with global climate change.

[5] "Getting the Best from Cities: Mobilizing Dispersed Interests to Anticipate Problems—Preventing and Managing Disasters" in Sustainable Development in a Dynamic World: Transforming Institutions, Growth, and Quality of Life (Washington, DC: World Bank and Oxford University Press, 2002).

[6] Ibid.

[7] Ibid.

[8] World Population Prospects, The 2002 Revision, ESA/P/WP. 180 (United Nations, New York, NY, February 26, 2003), p. 1.

[9] I (environmental impact) = P (population) · A (per capita affluence) · T (technology index).

[10] Tomorrow's Markets, Global Trends and Their Implications for Business (Washington, DC, World Resources Institute, United Nations Environment Programme, World Business Council for Sustainable Development, 2002), 22–23.

[11] Ray Anderson, Mid-Course Correction, 19.

[12] "Achievements and Challenges. The Core Development Challenge" in Sustainable Development in a Dynamic World: Transforming Institutions, Growth, and Quality of Life (Washington, DC: World Bank and Oxford University Press, 2002).

[13] Gregg Easterbrook, "Forgotten Benefactor of Humanity," Atlantic Monthly, January 1997, 75–82.

[14] Moore's law is named after an observation made by Intel's Gordon Moore. In 1965, while graphing memory chip performance data in preparation for a speech, he noticed a striking trend: each new chip contained about twice the capacity of the previous chip released 18–24 months prior. This trend has continued for almost four decades, without any end in sight.

[15] Lew Platt, "Rules of Survival in the Brave New World of Total Connectivity" (remarks made before the European IT Forum, September 7, 1998).

[16] Kevin Kelly, New Rules for the New Economy (New York: Penguin Books, 1999).

[17] Howard Rheingold, Smart Mobs: The Next Social Revolution (Cambridge, MA: Perseus Publishing, 2002). Reported in World Future Society's Future Survey 24, no.11, November 2002: 21.

[18] "Members Give Their Views on the Business Case for Sustainable Development," WBCSD News, October 17, 2003.

[19] Daniel C. Esty and Michael E. Porter, "Measuring National Environmental Performance and Its Determinants," The Global Competitiveness Report 2000 (New York: The World Economic Forum, Oxford University Press, 2000), 60–69.

[20] Forest L. Reinhardt, "Bringing the Environment Down to Earth," Harvard Business Review 77, no. 4 (July–August 1999): 149–157.

[21] Stuart L. Hart and Clayton M. Christensen, "The Great Leap: Driving Innovation from the Base of the Global Pyramid," MIT Sloan Management Review 44, no.1 (Fall 2002): 51–56.

[22] Alexander Barkawi, "Sustainability Investing—The Strategy for Long-term Shareholder Value," Sustainable Development International (Spring 2002): 51–52.

[23] Dow Jones Sustainability Indexes (press release, Zurich, Switzerland, September 4, 2003).

[24] Earnest O. Robbins II, "Sustainable Development Policy" (memorandum from the civil engineer, HQ USAF/ILE, DCS/Installations & Logistics, December 19, 2001).

[25] Dwight A. Beranek, "Sustainable Design for Military Facilities," Technical Letter No. 1110-3-491 (letter, Engineering and Construction Division, Directorate of Civil Works, U.S. Army Corps of Engineers, May 1, 2001).

[26] Arup's SPeAR model is discussed in detail in chapter 6.

[27] See Forest Reinhardt's article "Bringing the Environment Down to Earth," referenced earlier in this chapter (see n. 20).

CHAPTER 4

Walking the Talk

Our transformation to become a sustainable growth company is more fundamental than reducing our footprint and being socially responsible. The transformation is about using knowledge intensity and integrated science to develop products that will help make people's lives better, safer and easier.[1]

Between a rock and a hard place

If you are marketing engineering services in sustainable development to clients that have made a commitment to sustainability, be assured that your prospective clients will want to know your level of commitment. After all, your level of commitment is an indicator of your credibility and qualifications to work in this market. If you are not making a commitment to sustainability, those clients will question your ability to assist them in becoming a more sustainable organization since you apparently have not accepted (or perhaps do not fully understand) its foundations and principles.

Conversely, it is highly likely that many of your current and prospective clients do not understand, nor have they accepted, the precepts of sustainable development. In those cases, your statements about your commitment to sustainability principles may be perceived as a negative, reflecting a company that may simply be caught up in the environmental issue du jour. For U.S.-based companies, particularly those whose businesses are largely domestic, lack of understanding and acceptance of sustainable development precepts is more the rule than the exception. Furthermore, there are a small but significant number of clients who have taken what they have heard about sustainable development and judged it to be an overblown environmental concern, even antibusiness.

The purpose of this chapter is to provide the reader with a better understanding of what lies ahead in preparing his or her company for entering the market for sustainable development-related services. It is intended to answer, or at least to show the way for answering, a number of important questions about the prerequisites for market entry. What is your firm's business case for incorporating sustainable development into its corporate strategy? How do you ascertain whether your firm should make a public commitment to sustainability? Just how public should that commitment be? How should you go about greening your company, that is, making your company's operations more sustainable? How should you market sustainable development engineering services to clients?

Chad Holliday, Stephan Schmidheiny, and Philip Watts, three of the world's leading industrialists, argue that following the precepts of sustainable development is not only good for business but also essential for future economic growth and development.[2] In their book *Walking the Talk: The Business Case for Sustainable Development*, they offer many examples of how companies (mostly large multinationals) are incorporating the principles of sustainable development into their strategies and operations and improving both their top and bottom lines. Still, they admit that sustainable development is unknown to most businesses around the world. Further, they point out that many who know the term do not accept it or understand how it could affect their companies.

Engineering firms that intend to pursue this market need to carefully craft a business strategy that takes into account these differences in understanding and perception. The strategy should evolve from a critical assessment of the needs and interests of their clients, current and prospective. Care must be taken to create a strategy and public position that demonstrate to clients the firm's commitment to sustainability, while acknowledging the current level of understanding and controversy surrounding the issue.

Figure 4-1: Artist drawing of a hotel in Southern California. The design calls for solar, wind, and geothermal energy units; electric cars and trucks; environmentally benign building materials; and waste conversion units. Illustration courtesy of Paul Bierman-Lytle, Arrowhead Development Corporation.

Engineering firms should not try to preach sustainable development. Instead, they should approach the subject from the vantage point of their experience with many clients in multiple sectors of the economy. Each of these clients has a unique perspective on sustainable development issues. The firms' good working relationships with their clients have allowed them to see how issues of sustainability affect different industries and their respective stakeholders. Here, there are many lessons to be learned and shared. With this backdrop of experience, sustainable development can then be characterized as a confluence of economic, environmental, and social trends that is beginning to affect many aspects of industry and government operations. Although the impacts vary greatly among organizations, they appear to be fostering and inspiring new ways of doing business.

As an organization that has experience in working in a number of industry and government sectors, your firm can then present itself credibly as having broad knowledge about sustainability trends and issues and their potential impacts. This vantage point is an important asset, as industry and government organizations tend to focus mainly inside their own sector boundaries. With this knowledge, your firm can offer valuable assistance to clients in addressing these impacts and the associated risks and opportunities related to sustainable development.

Establishing a sustainable development practice

Establishing a successful sustainable development practice requires the following:

- Corporate commitment. Sustainable development principles have been endorsed by the engineering firm's leadership and incorporated into business strategies, policies, and programs, including public outreach. The degree to which the firm publicizes its commitment is a critical decision, based on client attitudes and disposition toward sustainable development.

- Resources, processes, and tools. The firm has acquired the necessary resources, processes, and tools to meet known and anticipated client needs. Much of the knowledge and skills required are usually not found in a traditional engineering firm.

- Continuous improvement. The firm is aware of how its own activities affect sustainability, and it has active programs for measurement, assessment, and improvement.

The scope and dimensions of each of these elements must be determined by the firm's leadership and integrated into its strategic plans and operations. The basis for this determination must be a well-thought-out business case for sustainability, one that aligns the firm's vision, mission, and goals to the needs and interests of its clients. Establishing the business case is a crucial step. The extent to which client organizations understand the issues and implications of sustainable development varies greatly, especially in the United States. Consequently, an engineering firm that fails to educate or misreads the interests of its clients runs the risk of losing its clients along with its credibility.

Clients expect firms that offer services in sustainable development to have not only subject matter expertise, but also to have incorporated sustainability principles into their corporate policy. Furthermore, they must have communicated those policies throughout the organization. A litmus test for clients will be the response of the firm's employees when asked about the firm's vision, policies, programs, and real commitment to sustainable development.

Strategy and policy choices: some tough decisions ahead

Making a commitment to the principles of sustainable development should not be treated lightly. The prospect of this commitment brings up a number of contentious policy issues and decisions that must be made regarding how your firm will run its business in the future. Some of the important issues and decisions follow.

Your first decision: is there a business case?

First and foremost, the major decision your firm needs to make is this: *What is your firm's business case for sustainable development?* It must make business sense for your firm to incorporate sustainable development policies and practices into its strategies and business plans. That is, your firm's commitment to sustainability must be seen as helping your firm meet its objectives for growth and profitability, market differentiation, recruitment and retaining of the best people, innovation, and otherwise increasing shareholder value. The assessment for and creation of a business case for sustainable development is discussed in depth later in this chapter.

A public corporate commitment to adhere to the principles of sustainable development begins at the highest levels of the firm. *It is crucial that the CEO and the board of directors understand the precepts of sustainable development and believe that a commitment to sustainability is directly connected to the firm's future success.* Furthermore, the CEO and his or her senior managers must convey that belief throughout the firm through words and actions.

Without such a consistent and credible message from the firm's leadership, the initiative will fail. Principles and policies regarding sustainability must be implemented firmwide. They cannot be adopted by just a part of the organization. Implementation can vary greatly among business units, based on the clients and markets those units serve. However, the basis of the commitment and the corresponding messages must be consistent throughout the organization.

The foundation for this level of commitment to sustainable development is the creation of a sound and compelling business case. *No matter how committed the CEO or high-level executives are to sustainable development, the commitment must be well aligned with the firm's business interests.* A firm and its leadership may believe in the precepts of sustainable development, but the corresponding strategies, policies, and practices must match the needs and interests of its clients, employees, and other key stakeholders. To reach that alignment, the firm must develop its own business case for sustainable development.

Building the business case

Creating a business case for a firmwide commitment to sustainable development starts with an assessment of the firm's business environment. The purpose of this assessment is to ascertain the needs and issues of key stakeholders as related to sustainability. The primary

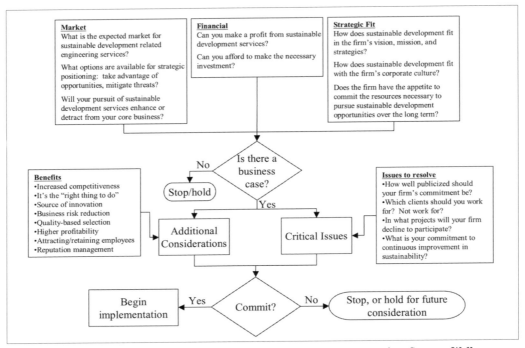

Figure 4-2: Decision process for establishing a sustainable development practice. Source: Wallace Futures Group, LLC.

stakeholders of the firm are its clients, employees, and competitors; however, other stakeholders—for instance, federal and state regulatory agencies or public interest groups—also may have significant influence on a firm's particular business.

This assessment must provide persuasive answers with good supporting documentation to the following questions:

Market

- What is the expected market for engineering services related to sustainable development, both current and future?

  o What are your market segments and how are the clients in those segments affected by sustainable development issues?

  o For the clients in each market segment, what are their needs and concerns regarding sustainable development?

  o What is the type and expected volume of services that your clients are expected to need related to sustainable development?

  o What is your firm's interest in and capacity to extend its market and geographic presence, now and in the long term?

- What options are available for strategically positioning your firm in the arena of sustainable development to take advantage of opportunities or mitigate threats?

- Will your pursuit of engineering services in sustainable development enhance or detract significantly from the firm's other core businesses?

Financial

- Can your firm offer and deliver the requisite services in sustainable development to its clients and make a profit?

- Can your firm afford to make the necessary investments in human resources, training and education, professional memberships, software tools, and references in order to build a sufficient sustainable development knowledge base?

Strategic fit

- How do the issues of sustainable development mesh with your firm's current vision, mission, and strategies?

- How does sustainability fit with your firm's corporate culture? Will this issue be embraced by the firm's managers and knowledge leaders?

- Can your firm pursue opportunities related to sustainable development effectively, that is, with the appetite to build the business and a willingness to commit resources over the long term?

The business case should be researched, assessed, and developed by a small task force of trusted individuals commissioned by the CEO and the board of directors. The task force should be supported by an executive sponsor, someone who can guide the group, ensuring that its work and deliverables meet the needs of the decision makers. Senior members of your firm should definitely be included in the task force but should not dominate the membership. The members of the task force should be considered "thought leaders" in the firm, people others listen to and follow. They should be recognized as having good business sense coupled with a strong track record of identifying new opportunities for the firm. In addition, it would be helpful to bring in someone equally qualified from outside the firm to provide a different perspective.

In making the business case, the final strategy choice depends primarily on your firm's client base, both current and prospective. The task force should work closely with your firm's market segment leaders and project managers to ascertain potential client interest and involvement in sustainable development. Some clients may already have incorporated sustainable development into their corporate strategies for the business reasons described in chapter 3. Others may be facing the same issues and market forces but may be unaware of the opportunities and threats.

Firms whose clients are based in the United States and have relatively small environmental footprints may find that sustainable development is not an important issue. Furthermore, some clients may actually react negatively to the concept, believing that the movement is the latest in a string of radical environmental issues. Conversely, firms whose clients are multinational companies or municipalities with sustainability policies may find that making a public commitment to sustainability and investing in the requisite resources will improve their prospects for project work.

A final note, which cannot be repeated too often: care must be taken to avoid the "evangelist trap." Taken outside the business context, sustainability is seen by many as an important and noble cause that is worthy of worldwide attention. As such, it is relatively easy for employees to become caught up in the cause and end up preaching sustainability to their clients. It is important in your internal messaging to remind employees that their role is to *advise* clients on the trends and potential impacts to their operations.

Benefits of integrating sustainability into the firm's corporate strategy

In addition to the opportunity for new business, there are additional benefits that accrue to firms that integrate sustainable development into their corporate strategy:

- *Increased competitiveness.* Firms that understand the issues surrounding sustainable development and can translate them into the needs of their clients will have a clear advantage over their competition. I argue that sustainable development is a "game-changing" issue. For firms that can identify and understand the business implications of the movement toward sustainability, the issue of sustainable development changes the basis of competition. Engineering firms that recognize the power of stakeholder opinion; the convergence of economic, environmental, and social issues; and the importance of environmentally friendly technologies, to name a few, will have a distinct advantage over firms that do not.

- *The "right thing to do."* If you and your firm understand and accept the assertion of sustainable development—that resources and ecological carrying capacity are not only finite but seriously at risk—then producing engineering designs and project approaches that follow the principles of eco-efficiency and social responsibility is the only ethical and responsible course of action. Even if you are not convinced about the risk, the high exposure levels suggest that it is smart to err on the side of caution.

- *Source of innovation.* As companies like 3M, Hewlett-Packard, and Kodak have found, sustainable development is a source of innovation and creativity. It encourages engineers to look at problems through different lenses and helps them recognize new opportunities, new approaches for engineering designs, and the need for new engineering services. Conversely, the issues of sustainability have exposed serious flaws in the logic and effectiveness of traditional engineering design approaches. Engineering firms that fail to recognize these problems may be seen as foolish or ignorant in the eyes of their clients.

Figure 4-3: Workers assembling a bridge in Crawford County, Kansas, using panels made of fiber reinforced polymer (FRP), which is five times lighter than conventional concrete and steel and has a much lower life cycle cost. The company, Kansas Structural Composites, Inc., is running tests on panels using recycled FRP. Photo courtesy of Kansas Structural Composites, Inc.

- *Business risk reduction.* Firms that recognize early the flaws in traditional design approaches exposed by sustainable development concepts will be able to alter their approaches sooner, thereby reducing their business risk.

- *Qualification-based selection.* Clients looking for professionals with knowledge and experience in sustainable development will make qualification-based selections, realizing that engineering firms that understand the principles and practices of sustainability often can produce work products that are more cost effective. For example, engineers who know how to apply life cycle analysis to the design can deliver a facility that has substantially lower operation and maintenance costs. Although the cost of the building may be higher initially, that added cost will be returned many times in terms of lower operating costs and higher levels of productivity.

- *Higher profitability.* The concept of sustainability is relatively new, and there are few, if any, practice standards or procedures to follow. As a result, clients who want to devise strategies or conduct projects related to sustainable development cannot do so by using conventional resources and commodity-priced labor. They will need experienced professionals with specialized knowledge and skills. Such resources are at a premium and can (and should) be priced accordingly.

- *Attracting and retaining employees.* Sustainable development is an area of high interest to young engineers and scientists because it offers a meaningful career opportunity—an opportunity to make a difference in the world. An engineering firm that has an active practice in sustainable development will be able to attract and retain the best talent.

- *Reputation management.* A commitment to sustainability and its precepts is a commitment to integrity, credibility, transparency, and accountability. Credibly demonstrating your firm's adherence to these principles will contribute greatly to the reputation of your firm.

Long-term opportunities in the developing world

The field of sustainable development offers a huge set of engineering opportunities throughout the United States as well as the rest of the world. Over the next several decades, World 1 nations will be engaged to varying degrees in changing out the current set of processes and technologies and replacing them with new systems more compatible with sustainable development principles. This change will require an enormous investment in infrastructure. If properly prepared and positioned, engineering firms can become deeply involved, inventing new processes and technologies, measuring and verifying their performance, and integrating them effectively into other processes and systems.

Furthermore, there is a greater potential demand for sustainable development engineering services in the World 2 nations. Their rates of population and economic

Figure 4-4: Students from Engineers Without Borders-USA working together with the people of San Pablo, Belize, on a project to bring water from the river to the village. Photo courtesy of Engineers Without Borders–USA.

Figure 4-5: Bringing fresh water to the village of Bir Moghrein, Mauritania. Photo courtesy of Engineers Without borders–USA.

growth are predicted to be much larger than those for World 1 nations and will result in increasing demands for goods and services. The demand for sustainable development engineering services depends on the extent to which these nations will employ sustainable technologies and processes to fulfill those increasing demands. Indications are that the World 2 nations are inclined toward sustainable development. As observed in chapter 3, both World 2 and World 3 nations are, in many cases, making sustainability a prerequisite for market entry.

World 3, the impoverished nations, need to meet basic human needs in agriculture, water resources, and sanitation. They need massive amounts of humanitarian aid just to survive. At the 2002 World Summit on Sustainable Development, sustainable development was reaffirmed as the central element in a global agenda to fight poverty and protect the environment. Furthermore, many of the World 1 countries pledged significant amounts of aid to other countries in need and made concrete commitments in the areas of poverty alleviation, clean water and sanitation, sustainable production and consumption, energy, chemicals, management of the natural resource base, health, and corporate responsibility. The importance of these commitments is that the World 1 nations pledged to commit significant amounts of monies toward meeting these goals within specific time frames. This translates into a significant number of projects in Worlds 2 and 3, projects that need the talents and expertise of the engineering firms.

If the World 1 nations are true to their commitments, these efforts will open up new opportunities in new geographies to engineering firms around the world. They will have new opportunities to bring their services to the other 5 billion people in the world. In addition to engineering services, the engineering community can help to improve the framework conditions necessary to build national economies. These are the fundamental steps toward poverty alleviation and building sound governments.

Critical issues

Related to your firm's overall business case for sustainable development are a set of controversial issues that must be addressed:

- *Just how well publicized should your commitment to sustainability be?* Firms that want to compete for sustainable development work may not want to directly advertise their commitment to, and actions taken in support of, the principles of sustainability. Others may want to announce their commitment as part of their overall strategic initiative strategy, telling their stakeholders about the sort of firm they intend to be. Either way, firms should recognize that public announcements along these lines tend to draw out knowledgeable critics who will want to assess your claims of commitment. Here, firms have a broad set of choices, ranging from little or no publicity (some firms call this "getting caught doing the right thing") to making sustainability the centerpiece of their marketing programs. The stronger the declaration, the higher the intensity of scrutiny.

- *Which clients should you work for?* Making a commitment to the precepts of sustainable development suggests to some of your stakeholders that there are certain clients that, because of what they do or the corporate policies they follow, are inappropriate clients for firms that have made a commitment to supporting sustainable development policies. For example, weapons manufacturers or nuclear power utilities are seen by some groups as engaging in the delivery of goods and services that are inherently nonsustainable. Other clients may have a poor track record in compliance with environmental laws. Still others may have been accused of having fraudulent accounting practices. Here again there are numerous critics—some quite knowledgeable but often with very narrow agendas—who will judge your commitment to sustainability in terms of which clients you choose to work for.

The engineering firm should not try to be all things to all people. Instead, it should develop a policy position that takes into account the current level of understanding and acceptance of sustainable development. The firm should seek to assist clients in forming a better un-derstanding of sustainable development on *their* terms, perhaps helping them develop their own business cases for sustainable development. Reaching a condition of sustainable development involves a journey, moving away from a vast array of current nonsustainable practices. Making judgments about current businesses or practices is neither helpful nor productive.

Figure 4-6: Cartoon from the Web site of the public interest group FloodWallStreet, connecting Morgan Stanley to problems of Three Gorges Dam. Courtesy of Curtiss Calleo.

- *What projects will your firm decline to pursue or participate in?* Similarly, your stakeholders and critics may believe that there are some projects that firms committed to sustainability should not take on. The largest example is China's Three Gorges Dam, proposed to be the largest hydroelectric dam in the world. This project has been widely criticized by environmental groups across the world for its displacement of almost 2 million people. Furthermore, the dam will inundate extensive areas containing chemical and radiological wastes, potentially mobilizing and releasing those wastes into the environment. Morgan Stanley Dean Witter is responsible for financing the dam and, consequently, is being heavily criticized. Organizations such as FloodWallStreet are criticizing Morgan Stanley's participation and have launched an Internet campaign to convince people not to sign up for Morgan Stanley's financial product, the Discover Card.

Your firm should consider its pursuit of and participation in projects on a case-by-case basis. As a practical matter, it will be difficult to judge which projects will be criticized and whether your participation will be seen by your stakeholders as a significant negative. The best course of action will be to establish internal project screening policies and procedures that conform to and support your overall commitment to sustainable development, while recognizing that your firm is in the project delivery business. Projects that clearly dismiss important sustainability considerations (e.g., loss of key environmental resources or communities) or are likely to create great consternation among key stakeholders should be avoided.

- *What is your commitment to continuous improvement in sustainability?* If a company makes a commitment to sustainable development, there is an implied obligation to assess its own operations, look for shortfalls, and institute projects to bring its environmental and social performance up to expected levels. Sustainability performance metrics specific to the individual firm and its industry sector are established. Performance is measured by commonly agreed-upon indicators plus consultation with stakeholders. Results are publicly reported, usually on an annual basis.

Figure 4-7: Construction work under way. The European Commission reports that the construction, operation, and demolition of facilities accounts for 40 percent of energy use and greenhouse gas emissions. Source: ClipArt.com and JupiterImages.

Since their biggest sustainability impacts are in paper consumption and travel, engineering firms may see their internal operations as having little overall impact. However, the designs they create for clients and methods by which projects are constructed often have substantial impact. For example, the construction industry in the United Kingdom reports that about 17 percent of wastes going to land disposal are directly related to construction activities. It is estimated that this industry uses 6 metric tons of material per person per year.[4] The European Commission reports that the construction, operation, and demolition of facilities accounts for approximately 40 percent of energy use and greenhouse gas emissions.[5] Accordingly, institutions have begun to address the issue of sustainable construction. To improve competitiveness in the construction and construction products industries, the European Commission is encouraging industry action to promote sustainable construction practices as a key part of competitiveness.

Furthermore, clients, particularly the large multinational corporations, may expect (or possibly direct) the engineering firms they hire to produce similar reports, or at least do the assessments. Sustainability reporting advocates like the Global Reporting Initiative (GRI)[6] see a company's impacts on the economy, environment, and society as extending beyond traditional corporate boundaries into their supply chain. Hence, the GRI sustainable reporting guidelines ask specifically for supplier performance.

The commitment decision

The decision by an engineering firm to commit, to table, or to dismiss a sustainable development initiative requires no special handling and should be channeled through the CEO and the board of directors in the same way other strategic initiative decisions are managed. However, since sustainable development may be seen as a new and radical strategic direction for the firm, additional validation is encouraged. To these ends, the CEO or other members of the firm's leadership should meet with clients to better understand the nature of their involvement with sustainability and how those issues affect their client's operations. It would also be useful to attend sustainability conferences to learn about current sustainability issues and talk with the experts in the field. The purpose of this engagement is for the firm's leadership to convince themselves of the reality and impact of sustainability trends and better understand any business opportunities and threats.

Although sustainable development may affect each business unit differently, the commitment to sustainability must be implemented firmwide. Sustainability strategies, policies, and programs must be integrated consistently across the firm, although each business unit should be allowed to design its own programs for marketing and project delivery, taking into account client and market differences.

The commitment decision should include a rollout plan, which takes into account any uncertainties about client needs, cultural fit, and difficulties in implementation. Pilot-scale implementation projects in one or more offices or business units are a way to integrate sustainability into the firm at a pace the organization can assimilate without compromising credibility. Lessons learned in the pilot phase can be applied in subsequent phases, leading to better acceptance.

Your next decision: how to set up the business

Incorporating sustainable development into the firm's strategic plan

The logical time to integrate sustainable development into your firm's strategic plans and programs is during the normal planning cycle. Most engineering firms create, or review and revise, their strategic plans annually using a strategic planning process customized to meet their leadership styles and the needs of the organization.

Most engineering firms have a strategic planning process. Usually this process involves a significant amount of research, meetings, and writing activities, culminating in a document intended to guide the firm toward growth and profitability. Although the work is well intended, it has been my observation that the resulting work products miss many of the fundamental elements of strategy.

The literature contains a wide variety of strategy frameworks, most of which contain some but not all of what I see as the required elements. Out of my research, I have devised a framework that defines three essential and distinct levels of strategy:

- *Corporate strategies.* These are the enterprise or firmwide strategies that define the overall nature of the firm: its purpose, objectives, boundaries, and approach to governance. Specifically, it "defines the range of business the company is to pursue, the kind of economic and human organization it is or intends to be, and the nature of the economic and noneconomic contribution it intends to make to its shareholders, employees, customers and communities."[7] Corporate strategies are revealed through statements of corporate vision, mission, and goals.

- *Competitive strategies.* These are strategies that define how the firm intends to compete in its chosen markets. For each market, competitive strategies specify the firm's direction for growth: adding new services and moving into new clients, new geographies, or both. They also specify for each market how the firm intends to differentiate itself and how it intends to fulfill priority client needs better than the competition. Tactical action plans detailing approaches to specific clients or market segments can be produced as needed.

- *Functional strategies.* These are strategies that define where and how the firm intends to invest in order to build strengths or shore up weaknesses. These investments map back to its competitive strategies for its chosen markets. Detailed plans and programs for implementing and monitoring these investments may also be developed.

Integrating sustainable development into your firm means incorporating those principles and practices into these three distinct levels of strategy.

Corporate strategies

Corporate strategy author George Sawyer further refines the concept of corporate strategy by breaking it down into three components: (1) foundation strategies that determine the character of the organization and set its boundaries, (2) choice strategies that are the criteria used to make resource allocation decisions, and (3) business integration strategies that, for complex businesses, speak to the integration of the various components. These components are described in table 4-1, along with examples of how sustainability might be incorporated.[8]

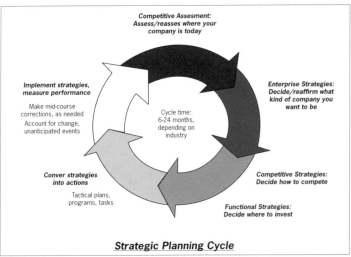

Figure 4-8: Strategic Planning cycle. Source; Wallace Futures Group, LLC.

This table should be a useful tool. By thinking through each of these components, the firm's leaders can construct the firm's policies on sustainable development.

Competitive strategy

While corporate strategies reveal your firm's overall purposes and objectives, competitive strategies show how the firm intends to compete in specific markets. Competitive strategies have three components:

*   *Strategic direction.* How the firm intends to grow revenue and market share in each of its markets. There are a number of distinct ways to increase market share:

    o   *Market penetration:* more intensive marketing of existing services to existing clients

    o   *Services diversification:* marketing new services to existing clients

Table 4-1: Components of corporate strategy

Strategy	Definition	Sustainability examples
Foundation strategies	States what sort of enterprise the firm wants to be and how management will direct the energies of the firms. Covers such topics as leadership, new opportunities, employees, communities, knowledge, and resources.	• Statement of commitment to sustainability principles • Effort and aggressiveness of pursuit of new opportunities in sustainable development • Employee reward and recognition: importance of employees as stakeholders • Strategies for working with the firm's communities as key stakeholders
Business integration strategies	Defines the relationship between different activities in the firm wherever these activities have strategic importance.	• Provision for drawing upon the resources of other business units in the firm to support multidiscipline sustainable development projects
Choice strategies	States the decision criteria used by management to allocate resources. Includes criteria such as diversification, manageability, harmony, quality of plans, and risk and reward.	• Degree of diversification into sustainable development work • Strategic fit of proposed sustainable development projects with the firm's more traditional projects • Extent to which sustainable development projects offer balance to the firm's business portfolio • Risk of sustainable development project success versus expected return on investment

o *Market diversification:* marketing existing services to new clients

o *Related diversification:* marketing new services to new clients, closely related to core businesses

o *Unrelated diversification:* marketing new services to new clients markedly different from core businesses

- *Approach.* How the firm intends to differentiate itself from its competition, that is, how it will meet the needs of its clients better than its competitors.

    o *Assessing the industry environment.* Making this determination begins with a sound assessment of your clients in each market segment: understanding their industry's environment, the competitive structure of their particular industry sector, and the trends and issues that affect their business. This assessment should conclude with a segment-by-segment appraisal of the clients' needs and wants.

    o *Making a competitor assessment.* The next step is to identify the firms that compete for your business in this market segment and assess their ability to meet important client needs. Knowing the ability of your competitors to meet the needs of current and prospective clients enables you to determine your strengths and weaknesses. A key point to remember is that a strength is only a strength if (1) your clients need and want that particular aspect of your service offering and (2) you can deliver that aspect better than your competitors. For weaknesses, the same logic applies.

    o *Determining your firm's approach for competitive differentiation.* Knowing your strengths and weaknesses enables you to determine the best approach for differentiating your firm in the market against your competition. Firms should pick two or three key areas of differentiation that seem to be long-term client needs in which your firm can continue to excel. Your selected areas of differentiation should be reexamined regularly to detect and respond to shifts in the clients' environment that may in turn affect their needs.

- *Method of development.* Overall investment (functional) strategies for building strengths and shoring up weaknesses. These range from internal development (internal education and selective outside hiring), to alliances (partnering and teaming with other organizations), to acquisitions of other firms.

Functional strategies

Functional strategies are specific investment strategies for building strengths and shoring up weaknesses, keyed to the functional elements of the firm. The list below offers seven functional areas common to most firms. Also included are the questions firms need to ask in order to prepare these strategies.

- *Marketing and business development.* What specific marketing and business development strategies will the firm use to gain a competitive edge? How will the firm maintain or expand our knowledge of the client base? Set pricing policies or strategies? How will the firm position with clients, develop relationships? How will the firm sell services in sustainable development?

- *Financial.* Are there any special long-term capital needs? Are there any large capital investments the firm needs to make to be competitive in this market? Should the firm consider making changes to our cost structure to accommodate special cost-reporting or pricing needs?

- *Human resources.* What strategic people resources will the firm need to acquire? Where should they be deployed? Are there special training, educational, and career development needs? Does the firm need to revise its reward and recognition policies to create a better environment for innovation?

- *Information technology and telecommunications.*[9] What are the hardware and software needs required to implement competitive strategies? What are the communications and network requirements? What software tools should the firm purchase or develop?

- *Operations.* What resources and capabilities do we need to maintain in order to support the competitive strategy? How should they be deployed? What tools, systems, and procedures for project delivery should we maintain? What are our standards of quality for this segment? How will we control the production systems? Track project progress? Manage workload leveling and staff utilization? Manage purchasing?

- *Services and technologies.* How will we identify and implement process improvements? Conduct competitor benchmarking? How will we determine client needs for new services? How will we track, maintain, or expand the firm's technology resource base? How much will we spend on research and development? What should our technology development portfolio look like?

- *Professional, governmental, and community relations.* What do we need to do at the federal, state, and local levels of government to support our competitive strategy? In what professional societies and trade associations should we participate, and what should our level of participation be? What other national, state, and local groups or nongovernmental organizations should we work with and develop relationships?

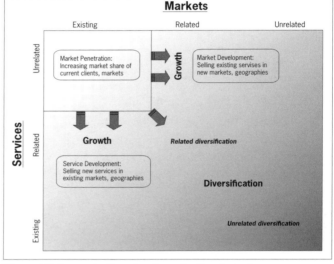

Figure 4-9: Strategic direction: how the firm intends to grow revenue and market share. Source: Wallace Futures Group, LLC

Organization

There are a variety of ways in which the firm can organize itself for marketing and delivering services related to sustainable development. The choice of organizational structure depends on the nature of the firm's client base, how the firm perceives the current market drivers for sustainability services, and the firm's readiness for delivering those services.

Establishing a sustainable development community of practice

Whatever the case, the important first step for any organization intending to market and deliver sustainable development services is to set up a community of practice in sustainable development that crosses all the business lines of the company. As an emerging field, sustainable development issues, experiences, and technological developments need to be identified, tracked, and shared throughout the organization. Although a firm may believe it has become quite knowledgeable about sustainable development, many of its clients have amassed an extensive sustainable development knowledge base as it pertains to their organization and their industry. Firms attempting to flaunt their sustainability knowledge for a specific industry run the risk of not knowing what they don't know. The engineering firm's *key strength* will be its broad knowledge about sustainable development across multiple business sectors, coupled with its ability to draw upon that broad knowledge and experience base for the benefit of its clients.

In knowledge management terms, a *community of practice* is a cluster of people organized around a common interest or area of knowledge. These groups capture, organize, manage, and maintain a body of knowledge that is used by community members as well as others throughout the organization to solve work problems. These communities also share experiences, insights, tools, and best practices, thereby increasing the value of the community to the organization. They also direct the flow of knowledge and promote consistent knowledge sharing throughout the organization.

The American Productivity & Quality Center (APQC) notes that five elements must be in place in order for communities of practice to do well:

- Endorsement and sponsorship by a senior member of the firm
- Defined membership
- Clear roles and responsibilities
- Accountability and measurement
- Supporting tools such as Web sites, "people-finders," and other collaborative software[10]

Establishing business units

In addition to a community of practice, a firm may want to establish one or more business units (or develop new service sets within existing business units) focused on delivering specific sustainable development services, such as green buildings, Leadership in Energy and Environmental Design (LEED) certification, life cycle assessment, and sustainable development reporting. Another option is to modify or recast some of its existing services, for example, energy savings, brownfield development, or urban and transportation planning, to incorporate sustainability. In either case, the organizational change should be made in close consultation with the key people in the business unit to ensure that the additional services or organizational change will be well received by current and potential clients.

Stewardship of the sustainable development community of practice poses an interesting dilemma. Logically, it would seem to fall within a business unit housing the environmental markets. However, given the broad applicability of sustainable development and the potential game-changing aspects of the business, it would be better to locate the practice, and perhaps some if not all of the new sustainability-related business units, under an entrepreneurial arm of the organization or at some location that encompasses all the relevant business units.

Scope of services

Table 4-2 provides examples of the scope of sustainable development services that firms might offer. Some of these services are new, while others are services that the firm has modified in order to incorporate elements of sustainability.

Knowledge and skill base

The creation of an effective knowledge and skill base to sell and deliver sustainable development projects begins with a strong foundation in the firm's traditional practices such as transportation, structures, buildings and facilities, power, soils and foundations, water and wastewater, air, and solid and hazardous waste. Using its detailed knowledge of its clients' business environment (customers, suppliers, and stakeholders), the firm can build its knowledge and skill base selectively and cost effectively by matching to current client needs and wants.

Preliminary staffing

Preliminary staffing of the sustainable development practice depends largely on the firm's perceived market for sustainable development services and on the level of corporate commitment. If the sustainable development market is expected to be strong and there is a solid commitment at the leadership level, then the firm should seek first to staff the sustainable development leadership position (business unit manager or practice director). Ideally, that person should come from outside the firm and have broad knowledge, experience, and reputation in sustainable development. Reputation is important for credibility with the client, to know that it is being served by a company that has made a serious commitment to sustainable development and seeks to bring high value to its clients. Such a person should be able

Table 4-2: Examples of sustainable development scope of services

Brownfield redevelopment	Industrial ecology
Building commissioning	Integrated water management
Carbon emissions trading	LEED certification
Compliance auditing	Life cycle analysis and planning
Corporate social responsibility	Material use management
Design for Environment (DfE)	Multimodal transportation systems
Ecological services assessment	Natural resource management
Energy auditing and management	Pollution prevention
Environmental cost accounting	Public, community involvement
Environmental impact assessments	Riparian corridor management
Environmental risk management	Smart-growth analysis
Greenhouse gas analysis	Sustainability gap analysis
Habitat restoration	Sustainability performance assessment
Health and safety	Sustainability policy development
	Sustainable development reporting

to discuss sustainability at the highest levels in the client's organization, hopefully developing project work on advising the client on sustainability strategies, policies, and programs. He or she can also assess the firm's current client and resource base and help determine what additional resources might be needed.

For small- to mid-sized firms, it may make sense to nominate someone from inside the firm to lead sustainable development activities. Ideally, this individual should be a mid- or senior-level person who can quickly get up to speed on sustainable development and upon whose business judgment senior management generally relies. That individual would be responsible for setting up and running any of the pilot projects deemed appropriate in the strategic planning sessions. He or she may also recruit a small internal group to track the sustainable development market and look to validate (or invalidate) the opportunities for the firm.

Table 4-3 lists examples of the types of skills that a sustainable development practice would draw upon, depending on client needs. The firm should have a cadre of engineers and scientists at all levels with a comprehensive knowledge of sustainable development. Some of the senior practitioners should have the ability to prepare and deliver papers to an audience knowledgeable in sustainable development.

To engage in sustainable development projects, the firm should be able to draw individuals from its resource pool with the types of experience listed below.

- Consult with senior-level officials in the client organization on matters of vision and strategy regarding sustainable development
- Run a public involvement program concerning sustainability

- Assess an organization's environmental impact
- Assess an organization's situation pertaining to corporate social responsibility
- Develop plans to implement sustainable development programs within an organization
- Analyze, assess, and develop a sustainable development report
- Design and implement a project that the client can defensibly claim contributes to sustainability

Developing stakeholder relations and partnerships

One of the most important ingredients for a successful sustainable development practice is the firm's relationship with stakeholders. This includes stakeholders of both the engineering firm and its clients. A major driving force behind the movement toward sustainability is the considerable power and influence wielded by stakeholder organizations. In a real sense, non-governmental organizations and citizen groups set de facto standards for corporate environmental and social performance. In effect, they give industry its license to operate. Therefore, the ability to communicate effectively with stakeholder groups and understand their needs and issues will contribute greatly to project success.

As they set up their sustainable development practices, engineering firms should identify early those stakeholder groups that affect their operations, both from a market and geographical standpoint. This exploration and identification should extend to the entire practices of the firm, since virtually all projects today could be affected—positively or negatively—by stakeholder groups.

Firms should consider developing partnerships with specific stakeholder groups. A working partnership with groups knowledgeable about particular subjects or locale could help the firm acquire essential knowledge and credibility. Likewise, sharing knowledge and experiences with the stakeholder groups could make for a better understanding of the issues in a particular venue.

Table 4-3: **Examples of skills for a sustainable development practice**

Environmental science	Planning
Climate change	Community planning
Environmental characterization	Livable communities
Environmental impact assessment	Multimodal transportation
Environmental justice	Stakeholder involvement
Environmental modeling	Resource efficiency: air, water, materials
Environmental policy	Urban sprawl, smart growth
Environmental processes	**Mechanical engineering**
Risk assessment	Energy efficiency in fluid flow
Sensors and measurements	Indoor air quality
Chemical engineering	Material selection
Bioprocess engineering	Renewable energy systems: wind, solar
Green chemistry	**Electrical engineering**
Environmental process engineering	Control systems
Hazardous chemical reduction	Distributed systems
Architecture	Renewable energy systems
Building reuse and retrofit	Sensors and measurement
Daylighting techniques	**Civil engineering**
Green building design	Bioremediation
High-performance building architecture	Energy audits
Construction management	Engineered wetlands
Construction material reuse	Materials selection
Deconstruction techniques	Natural hazards
Lean construction	Natural systems engineering
On-site pollution control	Water and wastewater treatment
Renovation, reuse of buildings	Water resources

Sustainable development tool kit

Below is a partial list of the tools that sustainable development practitioners would find useful. In addition, a number of useful papers, guidance documents, and software tools are provided on the accompanying CD-ROM. The contents of the CD-ROM are outlined in appendix B.

• Building commissioning	• Environmental impact analysis
• Building energy efficiency	• Green products sources, evaluation
• Chemical safety data sheets	• Evaluation of sustainability indicators
• Clean technologies assessment	• Greenhouse gas emissions reporting
• Community/stakeholder relations	• Life cycle assessment
• Corporate greening techniques	• Environmental and social assessment techniques
• Decision analysis	• Sustainable development reporting frameworks
• Design charrette techniques	• Sustainable development goals and indicators
• Ecoeffectiveness analysis	• Sustainability in buildings (LEED, SPiRiT)
• Ecoefficiency analysis	• Sustainability project checklist
• Environmental footprint analysis	• Waste reduction, recycling techniques

Activities, memberships in the community, and professional society and trade organizations

Traditionally, engineering firms have been active supporters and participants in professional and trade organizations, generally centered on their practices and markets. Organizations like the American Council of Engineering Companies (ACEC), ASFE (formerly the American Soils and Foundation Engineers), the National Society of Professional Engineers (NSPE), and the International Federation of Consulting Engineers (FIDIC) promote the business and professional aspects of consulting engineering. Engineering organizations such as the American Society of Civil Engineers (ASCE), the American Society of Mechanical Engineers (ASME), the Institute of Electrical and Electronics Engineers (IEEE), and the American Institute of Chemical Engineers (AIChE) support the technical side. Firms generally encourage their employees to join these engineering organizations to stay current with developments in their respective fields and to exchange knowledge and insights with their colleagues in other companies, including both clients and competitors. Many firms also encourage their employees to become active in the business side of engineering and to participate in organizations such as ACEC and ASFE at the local and national level. In addition to exchanging technical information, ACEC and ASFE provide a business voice for engineering companies, tracking and seeking to influence important issues, policies, and legislation. Their meetings, local and national, provide a forum for exchanges of ideas and information.

Many of these professional and trade organizations have become active regarding the sustainable development issue. Most are tracking the issue as an interesting development in their respective fields. Some have set up committees or task forces to provide their organization with a more systematic assessment of the impacts. While these may be very useful groups in which to participate, it is wise to keep in mind that they may be too parochial in their thinking to grasp all of the aspects of sustainable development.

Firms seeking to broaden their knowledge in sustainable development are encouraged to not only participate in their traditional professional and business organizations, but to investigate other organizations that have missions related to sustainable development. Sustainable development covers a wide range of issues and interests. Furthermore, at this early stage no single organization or discipline can claim extensive knowledge or influence. Therefore, it makes good sense for firms to collect information from many sources.

Greening the organization

Another important part of a firm's public demonstration of its commitment to sustainable development is its own active policies and programs to improve the sustainability of its own operations. For engineering firms, the direct impact of their operations on the environment is very small compared with their industrial clients. But, regardless of their impact, clients will expect engineering firms professing knowledge and experience in sustainable development to have leveraged that knowledge and experience to reduce their own environmental footprint. They would do so in order to be consistent with their stated sustainable development policies, enhance their reputation, and save money.

By making some relatively simple changes to their operations, firms can reduce waste, conserve resources, and reduce energy usage. Many of these changes will result in substantial cost savings. For example, replacing meetings requiring participant travel with teleconferencing can save thousands of dollars in avoided air and ground transportation costs. Employing automatic faucet shutoffs and proximity light switches can reduce electricity bills. Each of these changes will also demonstrate that the firm is taking an active and smart approach to resource conservation.

Greening programs also provide a way for employees in the firm to become directly involved, looking for ways to improve the firm's environmental, economic, and societal performance. Firms can set up pilot programs in one or more offices as a way of introducing sustainability to the organization. Pilot programs also provide a means to ascertain how sustainable development fits with the culture of the firm. A detailed discussion of the greening of the organization is provided in chapter 5.

Project delivery

Projects in which sustainable development is an important objective have a number of key differences over conventional projects. In general, these differences are brought about by the need to push beyond the limits of conventional technology in order to make progress toward sustainability. A detailed discussion of how to deliver a project that seeks to make progress toward sustainability is the subject of chapter 6.

Examples of these differences are highlighted in the following:

- *Increasing stakeholder involvement.* By their very nature, sustainable development projects will involve a different set of stakeholders, generally more in number and diversity. Time and resources for stakeholder engagement need to be incorporated into the scope of work.

- *Increasing requirements for systems integration.* Whereas a conventional project might be content to use the utility grids for its electricity, water, and wastewater services, a sustainable development project would likely incorporate recycling and reuse as well as other closed loop systems. This adds to the complexity of the project and increases the risk that the project might not meet client or stakeholder expectations. These risks and uncertainties need to be accounted for in the procurement specification and in the engineering firm selection criteria. They also need to be communicated to the stakeholders.

- *Using new technologies.* Further complicating the problem of systems integrations is the use of new, more-sustainable technologies. Projects seeking to make progress toward sustainable development will be using new and perhaps untried technologies. Just as in the case of systems integration, it is important to recognize and account for these uncertainties in the application of these technologies.

- *Going beyond compliance.* Frequently, sustainable projects seek to move environmental performance beyond the norms of regulatory compliance. Verifying this level of performance will be an important outcome of the project, as will sharing the results with other practitioners. To these ends, the scope of work might include the measurement, validation, and dissemination of the results.

- *Seeking to raise the bar.* Achieving sustainable development involves improving the eco-efficiency of the project's technologies and systems over the conventional. Since sustainable development is a journey involving continuous improvement, it makes sense to track the performance of other projects in terms of sustainability parameters and then seek to improve that performance over the previous project. Akin to going beyond compliance, the scope of work might also include accommodations for technology dissemination and transfer. It also should include procedures for revisiting and resetting project performance objectives. Procedures for setting new and higher project sustainability performance goals and measuring achievements are described in chapter 6.

For sustainable development projects, clients will tend to procure engineering service using qualification-based selection criteria. If price is the sole selection factor, it is highly likely that the engineer would opt for a conventional design over a design that would push the technology envelope. Conventional designs pose the least amount of risk since they incorporate "tried-and-true" systems and technologies. Conversely, a sustainable design would likely involve new and more complex systems, incorporating such elements as water recycling, natural lighting and ventilation, renewable energy systems, and more. Clients seeking to incorporate sustainability principles into their projects will need to select engineering firms that have the knowledge and willingness to incorporate nonconventional technologies and systems.

Regardless of the other project objectives, economic viability of a project can never be ignored. If a project does not make economic sense to the client, it will not be built.

Sustainable development project goals and indicators

Projects are starting to appear with sustainability as one of the declared objectives. However, to date most, if not all, of these efforts have been done without the benefit of any overall guidance or direction. Without such guidance and direction, these efforts, although well intended, may not produce much movement toward the goals of sustainable development. Worse, they may result in losing ground and place the reputations of both clients and the implementing engineering firms in jeopardy. For example, a project with the objective of reducing the amount of water used in a process might accomplish this by an overuse of energy or an increased use of toxic materials. In other words, the project may have achieved one sustainability goal, but only at the expense of others. If such a project is declared sustainable by the client or the engineer, both become open to criticism by their stakeholders and by society.

As noted above, the determination and application of sustainable development project goals and indicators is discussed at length in chapter 6.

Figure 4-10: Examples of corporate sustainable development reports.

Sustainable development performance reporting

Because of stakeholder pressures and organizational commitments, many companies are publishing annual environmental or sustainable development performance reports.[11] At this writing, there are no agreed-upon reporting standards, although several organizations, most notably the Global Reporting Initiative, are starting to develop them.

Most of these reporting initiatives are directed toward large multinational organizations with enormous potential environmental and social impacts. Many of these companies spend millions of dollars to assess their operations and prepare their reports. Many of the reports of multinational organizations can be viewed at the Web site of Sustainability-Reports.com.[12]

Useful guidance for environmental reporting can be found on the World Business Council for Sustainable Development (WBCSD) Web site on sustainable development reporting. Its latest report, *Sustainable Development Reporting: Striking the Balance*, can be downloaded from its Web site.[13] The site also contains information on the reporting practices of 23 companies. It provides a step-by-step guide for companies wishing to develop their own sustainable development reports.

To date, most of the engineering firms that have addressed their sustainable development performance have done so in terms of their operations, that is, paper usage, travel, office energy usage, community service, and employee equity. Since their environmental and social footprint is relatively small, they are not generally expected to produce a sustainable development report. However, things may change. Organizations such as the GRI are asking industry to report not only its own performance, but also that of its suppliers. Therefore, clients may require their vendors, including engineering firms, to report on their sustainability performance, regardless of their net environmental impact.

Looking strictly at its own direct impacts on the environment does not correctly consider an engineering firm's full role in the construction industry supply chain, as a designer and constructor of the built environment. As stated earlier in this chapter, the construction industry is a substantial user of energy and resources, perhaps larger than any other industrial sector.[14] Although it can be argued that the activities of engineering firms are directed by their clients, this may be seen by stakeholders as a distinction without a difference.

Credibility, transparency, and accountability

A commitment to sustainable development is made more credible to the stakeholders of the firm by making those commitments public. Client organizations such as the WBCSD and several trade associations including ACEC have developed policy statements on sustainable development. Engineering firms should consider following their clients' examples.

There are a number of instruments that a firm can use to demonstrate credibility, transparency, and accountability:

- *Value statements.* These include ethics, values, and principles. Examples are the CERES Principles,[15] the Hannover Principles,[16] the Natural Step (TNS) principles.[17]

- *Codes of conduct, pledges, and charters.* Examples include the Earth Charter,[18] the UN Global Compact,[19] and the Keidanren Global Environment Charter.[20,21]

These written and published statements need to be backed up with public statements by the leaders and senior managers of the firm.

Marketing and selling sustainable development services to clients

The organizations that have made a public commitment to sustainable development have done so after a long and arduous process. Their commitment means that sustainability has become part of their organization's business model. The fact that they have committed means that they have explored thoroughly the meaning of sustainability, its affect on their business, and its affect on their key stakeholders. Taken together, their commitment makes them very knowledgeable about all the issues of sustainable development as it pertains to their organization.

Approaching clients

Engineering firms that intend to offer sustainable development services to potential clients should be aware of this level of knowledge that these organizations possess. Before proceeding, firms should carefully evaluate the needs of these knowledgeable clients and be prepared to discuss their skills relative to those needs. Most often, these clients need the normal engineering services to be delivered by engineering firms that understand the issues of sustainable development and can add those considerations into the project without any additional education.

Organizations that are exploring sustainable development and are thinking about incorporating its principles into their operations represent an attractive opportunity for engineering firms. With the right knowledge and skill base in place, a firm can help these organizations work their way through the entire conversion, starting with strategic visioning and planning and continuing through projects, in order to make their operations more sustainable.

Here again, engineers should not try to preach sustainable development. Instead, they should approach the potential client from the vantage point of having a number of clients in many sectors of the economy. Sustainable development should be addressed as a number of economic, environmental, and social trends that, when pulled together, represent a new theme with many risks and opportunities. The issues of sustainable development affect each sector differently. The firm can assist that client in assessing the impacts, risks, and opportunities and in developing a strategy that best suits the client. This vantage point is an important asset, as client organizations tend to operate inside sector boundaries. A firm with broad knowledge about trends and issues affecting other sectors will be valuable to a prospective client.

It should be noted that engineering firms are typically *not the clients' firms of choice* when it comes to developing a sustainable development program. At this writing, clients do not view engineering firms as the resource for high-end management consulting services. Instead, they tend to use high-profile sustainable development consultants for the initial strategy work. These consultants bring a high level of credibility to the process, having developed a reputation for outside the box thinking.

Clients also seek out architects with sustainable development credentials to build new facilities. Buildings that have been certified as meeting certain sustainability criteria appear to be of high interest to companies that have made a commitment to sustainability. Perhaps this is because buildings are a symbol of what the company stands for.

Project screening

For engineering companies, the greater impact on sustainable development results from the client projects they deliver. For companies that have made a public commitment to sustainability, how their stakeholders view their selection and delivery of projects will affect substantially the credibility of their commitment. Therefore, it is vital that companies committed to sustainability understand the potential impacts of their projects across all the dimensions of sustainability. Furthermore, this raises an important company policy question: are there projects or clients that a company ought not to accept because that engagement might compromise their credibility?

> *To these ends, FIDIC has developed a sustainability checklist, a tool for the initial screening of projects and programs. This checklist contains a series of questions that relate to all three dimensions of sustainability—economic, social, and environmental. The checklist also covers a variety of project types and practices, including renewable resources, fishing and farming, hydraulic resources, infrastructure, industrial activities and technologies, extractive industries, urban and land development, building construction, and transportation.[22]*

Continuous improvement

By now it should be well understood that sustainable development is a journey and not a binary choice. Furthermore, the journey from where we are today to where we want to be will be one of continuous improvement, that is, the systematic improvement of the operations of our companies based on the availability of new, more-sustainable technologies and processes.

Programs for continuous improvement should contain the following components:

- *Baseline assessment.* The firm should determine its current environmental and societal impact. This should be done by first defining the scope of the firm's activities and assessing the impacts of those activities against existing performance standards, norms, or other benchmarked firms. Scope is a key decision in this assessment. The firm may want to confine its assessment to its internal operations or expand it to include its impacts during project construction or implementation. It may also consider incorporating the impacts of the projects themselves. Again, although it can be argued that the impacts of projects are in the control of the client, stakeholders may not see that distinction.

- *Set objectives for improvement.* Based on the results of the baseline assessment, the firm should devise a comprehensive set of objectives for improvement. The objectives should be measurable against established indicators. Schedules for meeting objectives should also be established commensurate with the firm's resources, client expectations, and competitor benchmarks.

- *Implementation.* Once objectives and schedules are set, the firm should devise programs for implementation and allocate sufficient funding to achieve the objectives. Regular sustainable performance reports should be generated and sent to top management. While the purpose of the reports is to compare results against plan, they should include some information on client expectations and competitor benchmarking in order to keep current with changing expectations. The reports should also contain an assessment of technology developments that could change current practices.

- *Review and revision.* The firm should schedule periodic reviews of the programs for continuous improvement in sustainability. The purpose of the reviews is to check progress against the objectives and plans and to see how program funds were spent. These reviews should be conducted annually as part of the firm's normal budget cycle. Based on program performance, client expectations, new benchmarks, new technologies, firmwide performance, or other variables, the objectives should be revisited and revised accordingly.

Since achieving sustainability is a long-term proposition, the firm should keep a good, year-to-year record of program performance. The purpose of this record is to show the results of a broad and continuous effort of the firm to move toward sustainability.

---

[1] Dawn Rittenhouse, business sustainability manager of DuPont, private conversation, February 2004.

[2] Charles O. Holliday Jr., Stephan Schmidheiny, and Philip Watts, Walking the Talk. "Chad" Holliday is the chairman and CEO of DuPont. Stephan Schmidheiny is chairman of Anova Holding AG. Philip Watts is the chairman of the Committee of Managing Directors of the Royal Dutch/Shell Group of Companies.

[3] Johannesburg Plan of Implementation, http://www.johannesburgsummit.org/html/documents/summit_docs/2009_keyoutcomes_commitments.doc.

[4] Construction Best Practices, "Sustainable Construction," Watford, UK, March 2003, www.cbpp.org.uk.

[5] The European Commission, "Sustainable Construction," May 2001, http://europa.eu.int/comm/enterprise/construction/suscon/finrepsus/sucop1.htm.

[6] The Global Reporting Initiative (GRI) is a UN-sponsored organization whose stated purpose is to build a consistent framework for reporting sustainable development performance. Its Web site can be found at www.globalreporting.org.

[7] Kenneth R. Andrews, The Concept of Corporate Strategy (Irwin, IL: Homewood, 1987), 13.

[8] George Sawyer, Business Policy and Strategic Management: Planning, Strategy and Action (New York: Harcourt Brace Jovanovich, 1990), 136–155.

[9] Information technology and telecommunications could be consolidated under "Services and Technologies" or "Operations." However, given the growing importance of this technology area, I believe it deserves separate consideration.

[10] Carla O'Dell, Stages of Implementation (Houston, TX: American Productivity and Quality Center, 2000), 30–32.

[11] The World Business Council for Sustainable Development, for example, requires each member company to produce an environmental or sustainable development performance report as a condition of membership.

[12] See Sustainability-Reports.com's Web site at www.enviroreporting.com.

[13] Sustainable Development Reporting: Striking the Balance, WBCSD, Geneva, Switzerland. June 2003.

[14] The European Commission, "Sustainable Construction."

[15] The CERES Principles can be found at http://www.ceres.org/our_work/principles.htm.

[16] The Hannover Principles can be found at http://myhero.com/hero.asp?hero=w_mcdonough.

[17] The Natural Step principles can be found at http://www.naturalstep.org/learn/principles.php.

[18] The Earth Charter can be found at http://www.earthcharter.org/.

[19] The UN Global Compact can be found at http://www.unglobalcompact.org/Portal/Default.asp.

[20] The Keidanren Global Environment Charter can be found at http://www.keidanren.or.jp/english/speech/spe001/s01001/s01b.html.

[21] An excellent discussion on corporate codes of conduct can be found in a paper by Rhys Jenkins, Corporate Codes of Conduct: Self-Regulation in a Global Economy, Paper No. 2, United Nations Research Institute for Social Development, Technology, Business and Society Programme, April 2001, http://www.natural-resources.org/minerals/CD/docs/other/jenkins.pdf.

[22] International Federation of Consulting Engineers (FIDIC), Business Guidelines for Sustainable Development in Consultancy Services (Appendix 3), (Geneva Switzerland, FIDIC, September 2002). Copies of these guidelines may be purchased through FIDIC, www.fidic.org.

CHAPTER 5

Greening the Engineering Company

When a company says "we're green," how can you tell if they're genuine, or how far they take it? I always lean towards positive encouragement, rather than criticism, and I know that the business world contains many people who sincerely want to make a difference. On the other hand, there are companies who want to persuade us that their disposable nappies are greener than the next ones, or that if you buy a particular product then, judging by the advertising, all the birds will sing, and Mother Nature will beam happily.[1]

Setting up a greening program

The purpose of chapter 5 is to assist engineering companies in the design and implementation of programs for corporate greening. *Greening* a company means to establish a set of programs aimed at reducing a company's environmental footprint and improving its contribution to its communities. Programs cover such areas as procurement (reduced use of paper, increased use of recyclable materials, or reduced use of toxic materials), travel (reduced amounts of air travel or increased carpooling and telecommuting), and energy savings (use of daylighting, automatic room lighting, and insulation). These programs can also be incorporated into the company's project work.

An extensive list of ideas for greening your company can be found in table 5-2. Although not the most important component of corporate sustainability, greening programs can make a persuasive statement about a company's commitment to sustainability to clients, employees, and other key stakeholders. If done correctly, greening activities can save money and, at the same time, enhance the company's image. In addition, employees can take home the greening ideas and techniques and thus benefit personally from the company's programs.

It is essential that the greening programs established by the firm fully support the firm's business objectives and strategies. Opportunities for greening are in ample supply, and it is easy for employees to become immersed in greening activities, sometimes at the expense of their assigned work. Therefore, it is important for the company to establish control over its greening programs, striking a balance between just doing good for good's sake and doing something that is productive, affordable, and makes strategic sense. Likewise, it is critical to connect a company's greening programs to its overall business case for sustainable development. Moreover, companies need to know their clients' attitudes toward sustainable development and how the company's greening programs are perceived.

Conversely, it is essential that the firm support the greening programs at the level to which management and employees agreed. To a large extent, the firm's greening programs make a contribution to the local community. Therefore, the firm's support of a greening program is,

Figure 5-1: Villagers, professional engineers, and students from the University of New Hampshire chapter of Engineers Without Borders–USA building a leach field in Santisuk, Thailand. Photo courtesy of Engineers Without Borders–USA.

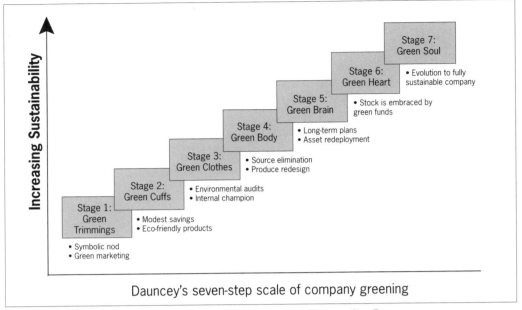

Dauncey's seven-step scale of company greening

Figure 5-2: Dauncey's seven-step scale of company greening. Source: Guy Dauncey, Sustainable Communities Consultancy, 395 Conway Rd, Victoria BC V8X 3X1, Canada; e-mail: guydauncey@earthfuture.com; web: www.earthfuture.com). Guy is the author of *Earthfuture – Stories from a Sustainable World*, New Society Publishers, November 1999.

in a sense, a public statement about the firm's support to its community. It also sends a signal to employees about the firm's true commitment to sustainable development.

As they develop and promote their greening programs, companies should review carefully any claims made about the company's greening program performance. Public declarations about sustainable performance will attract knowledgeable critics from environmental advocacy organizations, who will take great delight in debunking claims that are poorly conceived or inadequately documented. Consultant Guy Dauncey notes that Friends of the Earth, a United Kingdom environmental organization, makes an annual Green Con of the Year award to the company that makes overreaching claims of environmental goodness. For example, in 1999 the organization awarded the Green Con to the United Kingdom's Energy Saving Trust for allowing power companies to claim that their green energy products came from genuinely renewable resources. The group pointed out that the trust's so-called renewable sources of energy were incinerators that burned rubbish composed of materials (e.g., plastic and paper) that could have been recycled.

Dauncey offers this opinion about the seven-step scale of company greening. Although his opinion may seem a bit extreme, it does provide useful insight into the thinking of environmental advocacy groups with which companies may be confronted.

> *In Britain, Friends of the Earth have established an annual "Green Con of the Year" award, which they give, with much publicity (and accompanying embarrassment) to a suitably deserving company. It works wonders, and would be a very useful approach elsewhere too, if non-profit citizens' groups feel inspired to take it on. "My advice is be patient—and gently encourage a company to see if it can do more"*[2]

One of the challenges in building both interest and momentum in company greening activities has to do with perspective. How do you make employees aware of the impacts of company activities, indeed their own activities, on the environment? Mathis Wackernagel and William Rees have devised a useful concept—Ecological Footprint Analysis—that

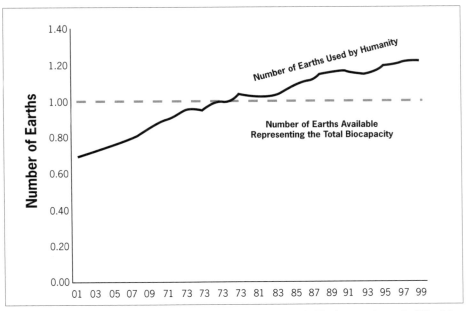

Figure 5-3: Ecological footprint: capacity of the earth. Source: Mathis Wackernagel, et. al., "Tracking the ecological overshoot of the human economy," Proc Natl Acad Sci U S A. 2002 Jul 9;99(14):9266-71. Copyright 1999 National Academy of Sciences, U.S.A.

converts the energy and resource flows of any economic unit (countries, companies, individuals) into the corresponding land and water area required to support those flows. It is a measure of the biological productive area needed to generate the resources and absorb the wastes of a given population. There are tools available for calculating ecological footprints for individuals and households. Employees can use these tools to get a sense of how their everyday activities affect the environment. A simple ecological footprint quiz is available on the Internet.[3] A more-detailed ecological footprint calculation for households is also available on the attached CD-ROM and on the Internet.[4]

Rigorous ecological footprint calculations are being done using the best available scientific data. Using these calculations, it turns out that the average person in the world needs 5.6 acres of biological productive area, while the average person in the United States needs 24 acres. Even more disturbing is the extrapolation to the global level. Given that there are about 4.7 acres per person available in the world, human activities are now exceeding the world's ecological capacity by about 20 percent.[5]

Client benchmarks

A useful perspective on corporate greening can be obtained by benchmarking clients and other organizations that have made commitments to sustainability. Each of the following organizations has made a public commitment to sustainability based on its own business case.

STMicroelectronics

This report reflects the role our culture and beliefs play in our commitment to build a successful company while contributing to sustainable development. Our culture is based on our integrity, the central importance of our people and our values.[6]

With annual revenues of $6.32 billion, STMicroelectronics (STM) is one of the world's top five semiconductor manufacturers. The company manufacturers over 3,000 products for

Table 5-1: STMicroelectronics Environmental Decalogue

Decalogue: STMicroelectronics's 10 Eco-mmandments[7]

1 Regulations
1.1 Meet the most stringent environmental regulations of any country in which we operate, at all of our locations.
1.2 Comply with all international protocols at least one year ahead of official deadlines at all our locations.

2 Conservation
2.1 Energy - Reduce total energy consumption (Kwh per k$) by at least 5% per year, through process and facilities optimization, conservation and building design.
2.2 Water consumption - continue to reduce water drawdown (cubic meters per k$) by at least 5% per year, through conservation, process optimization and recycling.
2.3 Water recycling - reach a minimum of 90% recycling ratio in 2 pilot sites by end 2005.
2.4 Trees - reduce office and manufacturing paper consumption (kg per employee) by at least 10% per year, and use at least 95% recycled paper, or paper produced from environmentally certified forests.

3 Greenhouse Gas Emissions
3.1 CO_2 - reduce total emissions due to our energy consumption (tons of carbon equivalent per M$) by at least a factor of 10 in 2010 versus 1990, which is a goal 5 times better than the average of the industries meeting the Kyoto Protocol goal.
3.2 Renewable energies - increase their utilization (wind, photovoltaics and thermal solar) so that they represent at least 5% of our total energy supplies by end 2010.
3.3 Alternative energies - adopt, wherever possible, alternative energy sources such as cogeneration and fuel cells.
3.4 Carbon sequestration - compensate the remaining CO_2 emissions due to our energy consumption through reforestation or other means aiming at total neutrality towards the environment by 2010.
3.5 PFC - reduce emissions of PFC (tons of carbon equivalent per M$*) by at least a factor of 10 in 2008 versus 1995.

4 Pollution
4.1 Noise - meet a "noise-to-neighbors" below 60dB(A) at any point and any time outside our property perimeter for all sites, or comply with local regulations (whichever the most restrictive).
4.2 Contaminants - handle, store and dispose of all potential contaminants and hazardous substances at all sites, in a manner to meet or exceed the strictest environmental standards of any community in which we operate.
4.3 ODS - phase out all remaining Class 1 ODS included also in closed loops of small equipments before end 2001.

5 Chemicals
5.1 Reduce the consumption of the 6 most relevant chemicals by at least 5% per year (tons per M$*), through process optimization and recycling (baseline 1998).

6 Waste
6.1 Landfill - reduce the amount of landfilled waste below 5% of our total waste by 2005.
6.2 Reuse or recycle at least 80% of our manufacturing and packing waste by end 1999, and 95% by end 2005.
6.1 Use the "Ladder Concept" as a guideline for all our actions in waste management.

7 Products and Processes
7.1 Design products for decreased energy consumption and for enablement of more energy efficient applications.
7.2 Contribute to global environmental control by establishing a database of Life Cycle Assessment of our products.
7.3 Systematically include the environmental impact study in our development process.
7.4 Publish and update information about the chemical content of our products.

8 Proactivity
8.1 Support local initiatives for sponsoring environmental projects at each site in which we operate.
8.2 Sponsor an annual Corporate Environmental Day, and encourage similar initiatives in each site.
8.3 Encourage our people to lead/participate in environmental committees, symposia, "watch-dog" groups etc.
8.4 Include an "Environmental Awareness" training course in the ST University curriculum and offer it to suppliers and customers.
8.5 Strongly encourage our suppliers and subcontractors to be EMAS validated or ISO 14001 certified, and assist them through training, support and auditing. At least 80% of our key suppliers should be certified by end 2001.

9 Measurement
9.1 Continuously monitor our progress, including periodic audits of all our sites worldwide.
9.2 Cooperate with international organizations to define and to implement eco-efficiency indicators.
9.3 Measure progress and achievements using 1994 as a baseline (where applicable) and publish our results in our annual Corporate Environmental Report.

10 Validation
10.1 Maintain the ISO 14001 certification and EMAS validation of all our sites worldwide.
10.2 Certify new sites within 18 months of their operational start-up, including regional warehouses.

more than 1,500 customers, including Bosch, DaimlerChrysler, Ford, Hewlett-Packard, IBM, Nokia, and Sony. Today STM has 33,000 employees dispersed in over 150 manufacturing, research and development, and sales offices located in 33 countries.

Business Week credits CEO Pasquale Pistorio with the company's surprising success. Formed out of a merger of two ailing microelectronics companies in the late 1980s, Pistorio guided the company to a number three ranking in market share. Only Intel and Toshiba are higher.[8]

STM has a strong commitment to sustainable development, and its commitment to responsible care of the environment has resulted in multiple prestigious awards and substantial cost savings. Energy and water conservation measures save the company about $50 million annually.[9]

> *For us, respecting nature is a precise business strategy, reflecting our many years of experience and conviction that environmentally aware companies are intrinsically more competitive and generate higher returns for shareholders.*

> *Our approach to the environment extends well beyond our Company, as we articulate clear sustainability criteria to our suppliers, significantly expanding our impact.* [10]

STM has adopted the International Chamber of Commerce's Business Charter for Sustainable Development as its operating guideline. In addition, the company issued what it calls the Environmental Decalogue: its vision for environmental management and sustainable development. In it, STM lists 10 "environmental commandments" covering such topics as regulations, conservation, greenhouse gas emissions, pollution, and waste. Each area has a subset of measurable environmental and resource conservation goals. The company's vision statement notes that the company strives for "environmental neutrality" or a zero environmental footprint.[11]

U.S. Air Force

> *It is Air Force policy to apply sustainable development concepts in the planning, design, construction, environmental management, maintenance, and disposal of infrastructure projects, consistent with budget and mission requirements.*

> *According to the Federal Acquisition Regulations (FAR), consultants for planning, environmental, design, and related professional services shall be selected partially on the basis of their "specialized experience and technical competence in the type of work required, including, where appropriate, experience in energy conservation, pollution prevention, waste reduction and the use of recovered materials."*[12]

In his presentation to the American Planning Association at its meeting in March 2001, Air Force senior planner Roger Blevins posed the question that was clearly on the minds of the audience: why is the Department of Defense (DOD) concerned about sustainability? As is turns out, the DOD has a number of reasons to be concerned. First, the DOD manages 25 million acres of land and believes strongly that it is responsible for the proper stewardship of that land. In that role, the department is becoming increasingly concerned about loss of resources and damage to ecosystems. Furthermore, growth and development—translated to urban sprawl and encroachment—is impinging on the mission capabilities of its installations.

In concert with the DOD policies and to better accomplish its mission and conserve resources, the Air Force has incorporated the principles of sustainability into its decisions and actions regarding the planning, design, construction, operation, maintenance, and decommissioning of its facilities. This has resulted in the development of sustainability guidelines, the testing and use of sustainable technologies, and the use of sustainability rating systems for its buildings and facilities. Furthermore, it has translated those principles into design principles and selection criteria in its engineering procurements.

The Air Force's sustainable development policy notes that beginning in fiscal year 2004, at least 20 percent of each Major Command project should be selected as LEED pilot projects. The ultimate goal for fiscal year 2009 is to have all military construction projects be capable of achieving LEED certification.[13]

City of Austin, Texas

Capital Improvement Program (CIP) projects must be coordinated and synchronized within a sustainable community framework and in a regional context. These major public expenditures are opportunities to reinforce the community's vision and serve as catalysts for sustainable strategies. Three general areas for a CIP strategic plan are presented—location of projects, financing, and a sustainability-based decision matrix.[14]

Like many cities, the City of Austin, Texas, is faced with many challenges of growth and economic development. To deal with these challenges, the city created a program called the Austin Sustainable Community Initiative (SCI). Since 1996, this program has been guiding the decisions of Austin, working in such areas as capital improvements, sustainable buildings, city department sustainability assessments, neighborhood planning, clean air, and smart growth. This initiative ties into Austin's own sustainable development action plan, to which the city has developed sustainability indicators specific to Austin and its criteria for community quality of life.

To evaluate alternatives and set priorities for development, the city has created its own capital improvement project evaluation matrix called the CIP Sustainability matrix.[15] The matrix

Criterion and Impact Indicators	

• Public Health/Safety
  o Public health
  o Safety
  o Crime prevention
• Maintenance
  o Maintenance
  o Protects assets
• Socio-economic Impact
  o Local job creation
  o Consider long term (not construction
  o Consider spin-off (not city jobs)
  o Job training
  o Public-private partnerships
• Neighborhood Impact
  o Preserves or adds heritage value
  o Adds or increases utilization or access
  o Increase property values
  o Adds/increases recreational opportunities
  o Adds/increases educational opportunities
• Social Justice
  o Equity
  o Diversity
  o Consider who is being served by project mission and location
• Alternative Funding
  o Grants
  o Aid
  o Bond alternatives
	• Coordination with Other Projects
	  o Coordination w/other Depts/projects
	  o Consolidation of services
	  o Synergy (Interconnect/Nexus)
	  o Shared operating system benefits
	• Land Use
	  o Regional sustainability
	  o Preservation of sensitive lands
	  o Adds new asset in nodal area
	  o Transit or Pedestrian Oriented Dev.
	  o Improves/increases carrying capacity of existing infrastructure in nodal area
	  o Is above in area slated for "nodal dev"
	• Air
	  o Zero-pollution
	  o Optimization
	• Water
	  o Zero-pollution
	  o Optimization
	  o Conservation
	• Energy
	  o Conservation
	  o Optimization
	  o Renewables
	• Biota
	  o Diversity
	  o Preservation
	  o Restoration
	  o Location
	• Other Environmental

Figure 5-4: City of Austin's Capital Improvement Project Evaluation Matrix Criteria and Impact Indicators. Source: City of Austin Capital Improvements Program (CIP) Sustainability Matrix: http://www. ci.austin.tx.us/sustainable/matrix.htm

works as a tool to help the city ascertain whether a project will contribute to the community's desired movement toward sustainability. Some of the objectives and strategies that would lead to a positive matrix score are the following:

- Investing in economically or socially disadvantaged areas of the region
- Maintaining and optimizing use of existing facilities
- Minimizing impact on ecologically sensitive areas
- Reducing sprawl and improving intermodal transportation access and use
- Reducing the heat island effect
- Reducing or eliminating emissions that are harmful to human health, ecosystem functioning, or climate stability
- Designing projects with aesthetic qualities and heritage value[16]

The Greening of Dana

As we considered the renovation we thought we should "walk the talk"—the building should be a place where principles of environmental responsibility are not only taught, but upheld and demonstrated to the community.[17]

The University of Michigan is renovating the 100-year-old S. T. Dana Building in a way that preserves its history while reaching for the highest levels of green building design and performance. It currently is the home of the School of Natural Resources and the Environment, the place where students learn environmental sustainability. The renovation is a four-year project incorporating all aspects of sustainability, not only in the design of the renovation but in the construction activities as well. Contractors are required to comply with stringent construction and demolition waste management practices, avoiding landfill disposal to the maximum

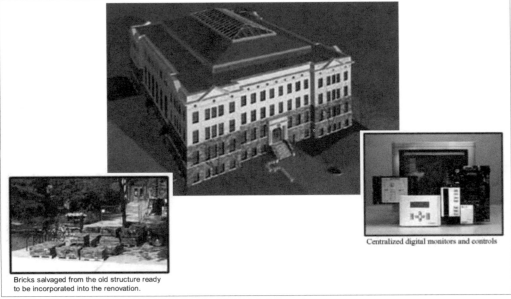

Centralized digital monitors and controls

Bricks salvaged from the old structure ready to be incorporated into the renovation.

Figure 5-5: Greening of the historic S. T. Dana Building at the University of Michigan. Extensive use of recycled and salvaged materials, renewable energy, energy and water conservation. The building resources and energy systems are constantly monitored and controlled by centralized digital systems. Many spaces can be tailored to individual needs. Source: University of Michigan, School of Natural Resources and Environment.

extent possible. In addition, the ductwork was sealed during construction to prevent dust and volatile organics from entering and contaminating the ventilation systems for years hence.

The building incorporates active and passive solar panels; composting toilets; water-saving plumbing fixtures; extra insulation; efficient lighting systems; and recyclable, renewable, and reusable materials.

Notably, the design team for the project includes the green architecture firm of William McDonough and Partners, the local architecture firm of Quinn Evans Architects, and the Netherlands-based green engineering firm of Ove Arup and Partners.

Alcoa

We believe that sustainable development is the basis for Alcoa's future. Within our company, we have the vision, the values, the strength, the technology, the people, the relationships, and the products to build that future. Aluminum—the core of our business—is one of the most useful, most easily recycled, most sustainable materials available in the world. Our customers provide our opportunity for growth as they challenge Alcoa to help them create solutions for improving the quality of life worldwide.[18]

Alcoa is the largest aluminum producer in the world, with 2002 revenues at $20.3 billion. It sees its product—aluminum—as being intrinsically sustainable. Of the 680 million tons of aluminum produced since the industry started in 1886, 440 million tons are still in use today. Yet despite its usefulness and recyclability, the processes and technologies for extracting and refining the raw materials and converting them to aluminum are resource and energy intensive. Furthermore, the process emissions and waste products are voluminous and relatively toxic.

In response, Alcoa has made a substantial commitment to sustainability. Because of its energy-intense processes, climate change is the defining issue for the company. Accordingly, Alcoa has decided to move beyond the debate over the science of climate change to addressing the needs of its stakeholders.

Industry must produce more and more to meet higher consumer demand while at the same time integrating social and environmental values into their operations. While there is little understanding within the community about the implications of these demands, the very real trade-offs involved, that lack of understanding does not absolve us from the obligation to manage these competing imperatives.[19]

Alcoa strives to be the best company in the world as judged by its stakeholders. It has set challenging environmental, health, and safety goals, including a 25 percent reduction in greenhouse gas emissions by 2010 or, assuming success on a new inert anode technology, a 50 percent reduction. This and other objectives are part of Alcoa's 2020 strategic framework for sustainability.

Establishing a greening program in your company

The most successful greening programs are established through a series of small steps taken in the form of pilot projects for an office, department, or other organizational unit. Firms of virtually any size can benefit from many of the greening ideas that follow. Starting this way allows the program to be assimilated into the company, building upon a string of small successes and learning experiences. As a result, the greening program evolves naturally, based on both successes achieved and lessons learned.

To establish a greening program, a company should follow the following steps. A list of ideas for greening the organization is provided in table 5.2.

1. *Create a small greening task force.* Set up a task force composed of a sponsor from senior management (executive sponsor) plus three to five people representing key parts of the

organization. The members of the task force should be considered thought leaders in the company, that is, respected for their ideas and contributions throughout the firm. The task force should be representative of the company in terms of age, gender, technical background, and management position. Initially the task force should have a small budget for meetings and communication materials. Having a budget is important as it represents a tangible commitment by the company to greening.

2. *Select candidate organizations in which to pilot greening projects.* The task force should investigate and select candidate organizations within the company in which to pilot a greening program. Candidate organizations must have a manager who is both interested in the program and is willing to sponsor and fund (at least in part) the cost of running the pilot project. In addition, the organization must have operations or practices that would benefit from a greening project and enable the task force to demonstrate an early success. The task force should select a mix of candidate organizations, such as an office, a practice, and an administrative department, in which to conduct the pilot projects.

3. *Identify candidate pilot greening projects.* Once the candidate organizations are selected, the task force should work with the organizations to identify pilot projects and generally define the scope, budget, and duration. For a typical company, five to seven candidate pilot projects would provide a reasonable pool from which to select. In general, projects should focus on ways to improve the eco-efficiency of the operation: reduce material intensity, reduce energy intensity, reduce toxics dispersion, enhance material recyclability, maximize sustainable use of renewables, extend product durability, and increase service intensity. In addition to the ideas provided in table 5.2, a good way to begin to identify projects is to engage the company's key stakeholders. This would entail one or more facilitated meetings in which task force members and others would conduct open discussions with key stakeholders and identify issues and priorities relative to sustainability.

4. *Select pilot projects for implementation.* From the candidate pilot projects, the task force should select two to three projects for implementation. Once selected, the task force should identify a project leader to run the project and further define the scope of work and budget. In addition, the project manager should also work with the task force to define measures of project success.

5. *Develop a project scope of work.* In crafting the project scope, it is important to contact affected functional departments within the organization. For example, if the scope of the pilot project focused on the greening of a regional office and sought to increase the use of paper with high recycled content, the pilot project leader should check with the person or organization responsible for paper purchasing to make sure this change in supply is compatible with the company's marketing and communication policies, equipment, budget, or other matters.

6. *Pilot project implementation.* Once the scope of work, schedule, budget, and measures of success are defined, the project should be ready to proceed. In the early stages, it is recommended that the pilots proceed without much advertising or fanfare. These efforts should be presented for what they are: pilot projects designed to examine the benefits of a corporate greening program.

7. *Measure and assess results.* As the project proceeds, the managers of the pilot projects, together with members of the task force, should measure and assess the results of the projects against the predetermined measures of success, especially cost savings.

At some point during and at the conclusion of the projects, all project managers and task force members should meet to discuss lessons learned across all projects. Knowledge gathered here would be valuable in directing the course of future greening projects and programs.

After reaching the end of the first set of pilot projects, the task force should meet with the executive sponsor plus other members of senior management to review the results and decide on the next steps. If the projects are successful, management may want to continue the projects, perhaps expanding them to other parts of the organization. Alternatively, management may want to modify or discontinue the effort based on lessons learned.

- Marketing and business development
  - o Develop brochures, proposals, qualification statements (SOQs) produced from recycled paper products.

Table 5-2: Table of ideas for corporate greening organized by functional area[20]

  - o Produce marketing and sales materials and displays that are flexible and reusable.
  - o Download marketing and sales publications from the Internet for just-in-time delivery.
  - o Cancel unwanted or redundant mailings, "junk" mailings received.
  - o Expand the use of electronic deliverables.
- Accounting and finance
  - o Use electronic invoicing and billing.
  - o Replace paper forms with electronic forms.
  - o Revise procurement processes.
    - Δ Buy paper that is equal to or greater than 30 percent recycled content, chlorine free, and from sustainably managed forests.
    - Δ Use vendors that employ sustainable practices.
    - Δ Choose products with Energy Star or other credible "ecolabeling."
    - Δ Do not purchase products that use excessive, nonreusable packaging.
    - Δ Purchase goods that are durable.
    - Δ Avoid purchasing single-use or disposal products.
    - Δ Support local producers.
    - Δ Seek to purchase products that are returnable to the vendor or the remanufacturer for recycling and reuse.
- Facilities and operations
  - o Revise printing and publishing processes.
    - Δ Use double-sided copying and printing.
    - Δ Use finer print and draft print formats for draft documents.
    - Δ Reduce the preparation of hard copy publications.
    - Δ Use lower-weight paper for internal use, drafts.
    - Δ Maximum use of electronic deliverables.
  - o Reuse and recycle supplies
    - Δ Replace paper cups with ceramic cups.
    - Δ Recycle paper, glass, plastic, cardboard, batteries, magazines, toner.
    - Δ Use supplies with high recycled content.
    - Δ Reuse in-house mailing envelopes.
    - Δ Establish programs to return and restock surplus office supplies.
    - Δ Fully use products to the end of their useful lives.

Table 5-2: Table of ideas for corporate greening organized by functional area (continued)

△ Provide recycling and waste collection areas throughout the facility.

o Revise furniture purchasing and use procedures

△ Purchase high-quality used furniture and recycle old furniture.

△ Use furniture that is flexible and adaptable to multiple configurations.

△ Order furniture that has high recycled content and low VOC emissions.

o Evaluate facilities location and site selection.

△ Locate near public transportation.

△ Preserve natural habitats.

△ Design the building to make the best use of natural shading, solar heating, daylighting access.

△ Protect significant wetlands.

△ Retain buffer areas.

△ Preserve natural drainage systems.

△ Account for prevailing winds for conditions favorable for passive cooling, natural shielding from adverse winds.

△ Concentrate development impacts on a small footprint while retaining open space.

o Evaluate landscaping.

△ Make appropriate plant selection: use native plants whenever possible.

△ Consider the allergy potential of plants.

△ Use turf grass in recreational areas only.

△ Evaluate nonnative plants to make sure they are noninvasive.

△ Employ xeriscaping whenever feasible.

△ Design irrigation systems to prevent overwatering.

△ Reduce dependence on limited or nonrenewable aquifers.

△ Incorporate a drought plan in the design.

△ Use integrated pest management techniques.

△ Use electronic sensors to determine watering requirements.

△ Use reclaimed water, rainwater, gray-water irrigation whenever possible.

△ Storm-water management

 − Use loose-set rocks, gravel, or porous paving to improve drainage.

 − Minimize impervious surfaces to control storm-water runoff.

 − Consider green roofs for storm-water management, temperature control in buildings.

 − Reduce pollutants in storm water by minimizing use of road salt and avoiding use of herbicides, pesticides, fertilizer.

o Evaluate facilities design.

△ Use low-VOC adhesives, paints.

△ Install automatic sensors on office lighting.

△ Use the maximum amount of daylighting.

△ Design the building to achieve LEED certification.

△ Use window coverings, awnings to reduce solar heating in summer.

△ Use drapes, shutters to control indoor temperatures.

△ Keep air registers, radiators free of obstructions.

△ Install ceiling fans.

△ Select light wall and ceiling colors to better reflect daylight.

Table 5-2: Table of ideas for corporate greening organized by functional area (continued)

- △ Seal heating and cooling ducts to reduce leakage.
- △ Perform regular maintenance on HVAC systems.
- △ Use rainwater collection systems.
- △ Use natural ventilation to the extent practical.
- △ Use light-colored roofing to reduce interior temperature.
- △ Use glazing systems that reduce energy use and enhance human comfort.
- △ Apply insulation based on life cycle cost analysis.
- △ Use wood harvested from certified sustainably managed forests.
- △ Use carpets made from materials that can be recycled.

o Establish equipment use procedures.
  - △ Turn off nonessential equipment each night.
  - △ Unplug or turn off unnecessary appliances each night.
  - △ Use task lighting.
  - △ Install high-efficiency lighting.
  - △ Install occupancy sensors for lighting in rooms.
  - △ Install programmable thermostats.
  - △ Adjust temperature settings to reduce energy costs.
  - △ Reduce hot water temperature settings.
  - △ Insulate hot water tanks.
  - △ Install automatic on-off controls on faucets.
  - △ Establish programs to report leaky faucets for repair.
  - △ Consider alternate forms of energy, e.g., wind, solar, biomass, use of waste heat.
  - △ Use or retrofit existing toilets with low-volume or waterless units.
  - △ Use or retrofit showerheads with low-flow units.

o Reduce transportation needs
  - △ Use hybrid cars in vehicle fleet.
  - △ Set up programs for telecommuting.
  - △ Encourage walking or cycling.
  - △ Encourage carpooling, use of public transportation.
  - △ Reduce air travel.
  - △ Replace trips with telecommunication, videoconferencing.
  - △ Provide shuttle service from public transportation stops to the office.
  - △ Provide pedestrian-friendly access to public transportation.
  - △ Provide safe, convenient facilities for bicycle commuters.

- Human Resources

  o Set up programs for telecommuting.
  o Stagger working hours to avoid rush hour commuting.
  o Provide incentives or "perks" for employees who use public or alternate forms of transportation.
  o Establish employee environmental awards.
  o Establish a greening suggestions program.
  o Set up Green Teams to develop ways to improve, reduce the firm's ecological footprint.
  o Provide greening training and education programs.
  o Maintain and distribute resource materials.
  o Provide educational materials, resources, and tools to employees for applying greening practices at home.

Table 5-2: Table of ideas for corporate greening organized by functional area (continued)

- Information technology and telecommunications
  - o Turn off computers, monitors each night.
  - o Work with the IT department to figure out ways to reduce energy usage.
  - o Make duplex printers the company standard.
  - o Facilitate access to electronic documents.
  - o Develop processes and technologies for telecommuting, teleconferencing.
- Professional and community service
  - o Become involved, sponsor sustainable development-related activities and projects in local communities.
  - o Educate, share information on sustainable development with professional societies, communities.
- Services and technologies
  - o Develop ways to track the application of greening techniques and determine cost savings.
  - o Internally publicize the company's green performance against goals, other benchmarks.
  - o Track the development of new greening technologies and systems that could save money, improve performance.
  - o Set up pilot programs to apply new technologies and systems and assess performance.
  - o Develop a "best practices" program to find, assess, and disseminate best practices throughout the company.

According to the seven-step scale offered by Guy Dauncey, firms employing the ideas listed in table 5-2 would still find themselves at stage two, a relatively low rank. Nevertheless, for firms that are in the beginning stages of a sustainable development program, an internal greening program is a good way to explore the cost and environmental benefits of becoming a sustainable company. It also makes a public statement about the firm's commitment to sustainability, to the firm's employees as well as its external stakeholders.

Alliances with other organizations

In addition to the greening ideas and examples in this chapter, companies should contact their respective engineering organizations to learn what sustainable development-related activities may be under way. For example, organizations such as the American Council of Engineering Companies (ACEC), the American Society of Civil Engineers (ASCE), the Institute of Electrical and Electronics Engineers (IEEE), the American Institute of Chemical Engineers (AIChE), the International Federation of Consulting Engineers (FIDIC), and the American Society of Mechanical Engineers (ASME) have issued policies on sustainable development and have assembled considerable information on the issues relevant to their members. Many are creating special committees of like-minded companies and individuals to share knowledge and experiences. Others have issued guidance documents and other publications that connect sustainability issues to the work of their members.

Other organizations outside the engineering profession offer valuable information on sustainability principles and practices. For example, the organization Business for Social Responsibility[21] helps companies establish and demonstrate their commitment to corporate social responsibility. It offers consulting expertise and training as well as numerous publications and opportunities for networking. In addition, the U.S. Business Council for Sustainable Development,[22] a regional arm of the WBCSD, is set up to develop and deliver projects in the United States that demonstrate the business value of a corporate commitment to sustainable development.

Finally, a new breed of humanitarian engineering organization is beginning to emerge. Formed in 2000, Engineers Without Borders–USA (EWB-USA)[23] is a nonprofit organization set up to assist developing communities around the world by delivering projects that help meet the basic human needs: safe drinking water, sanitation, health, energy, shelter, jobs, and education. The work is accomplished by both people in the community and engineering students working closely with their professors and professional engineers. EWB-USA's primary focus is to improve the lives of people in developing communities. However, the organization also seeks to develop a new kind of engineering student and career professional, an internationally knowledgeable and culturally sensitive individual with true "hands-on" engineering experience in the developing world. Seeing the value to their organizations, companies and government agencies as well as professional and trade organizations are forming alliances with EWB-USA and making donations to its projects and operations. The value they see is threefold: (1) providing a way for their membership to give meaningful assistance to developing communities in the form of labor and engineering know-how, (2) demonstrating a strong commitment to sustainability, and (3) improving the quality of engineering education.

Sustainable development reporting

As its greening program matures, your company may want to consider producing and publishing an environmental or sustainable development report. This report, generally produced annually, portrays your company's actions toward enhancing the environment as well as the communities in which it operates. The report could be in the form of a brochure illustrating your company's positive actions and contributions. A more rigorous version might contain a systematic assessment of the impacts on the environment and society along with a response and a commitment for improvement. For small- and medium-sized firms, the report could be part of a newsletter or be placed as a regular feature on their Web sites.

Right now engineering firms are not expected to produce rigorous analyses of their environmental and social impacts. That situation may change in the future for reasons described later in this chapter. But for now, the impacts from engineering firms are thought to be minimal when compared with the impacts of the large, resource-intensive and highly regulated companies. Nevertheless, it may be useful to develop some means to assess and communicate to clients, employees, and other key stakeholders the firm's commitments, activities, and accomplishments regarding sustainable development. Such reporting, if well integrated into a company's marketing and communication plans, can enhance a company's image and reputation.

In considering the development of environmental or sustainable development reports, the engineering firm must be careful. Whether it is a simple brochure or a detailed report, claims of contributions to sustainable development will attract criticism from green groups that have their own agendas and metrics for determining what constitutes good environmental and social performance. Oddly enough, companies with remarkably good records of environmental and social performance have found themselves the target of green groups for no other reason than that they made the performance information available for scrutiny. Meanwhile, these groups ignored the companies that produced no environmental or sustainability reports.

Companies desiring to produce some kind of communications regarding their environmental or sustainability performance should first meet with their stakeholders as described in the previous section to determine important information needs and issues. Such engagements will help frame the communication strategies and guide the development of the reports.

Environmental reports in various forms began appearing in the 1980s, coming mostly from the large, heavily regulated companies whose environmental impacts are substantial. The intent of this reporting was to relieve some of the pressures being applied by the environmental and public advocacy groups by presenting in a positive way their actions toward improving their environmental performance.[24] Recently, these reports have expanded in scope to include actions for social good. They have also changed in their communication approach, shifting from infomercials depicting their good works to detailed reports containing assessments of their environmental and social performance.

Starting with a commitment from the CEO, these reports seek to show that the company has taken a comprehensive look at its performance, has identified significant gaps, and has established reasonable goals and programs for improvement. In many of the reports, the companies have made considerable efforts to not only provide detailed information, but also have sought to communicate that information in a way that is nontechnical and very understandable. To enhance the credibility of the report, many of the reporting companies have sought to verify the information using outside firms to essentially audit the information. Since a sustainable development report is somewhat of an extension of a company's annual financial report, the audit work (and increasingly the work of data gathering and assessment for the sustainable development report itself) is given to the company's outside accounting firm.

Sustainable development reports are now being published on company Web sites, both to save the expense of publishing paper copies and to reduce paper usage and thereby avoid possible criticism. Having a company's story of its commitment to save resources printed in large numbers on high-gloss paper would seem ludicrous to the very audience the company is trying to reach. Still, many companies produce paper copies in an effort to reach a broad audience, many of whom do not have Internet access. The best approach seems to be a combination of both, a normally published version in combination with a Web-based version that provides up-to-date information on company performance. Often the Web site encourages visitors to answer survey questions, looking for feedback on the information provided as well as additional issues the report may have missed.

Demands for disclosures of environmental and social performance by individual corporations have been expanding. Bolstered by the recent scandals in corporate governance in companies like Enron and WorldCom, environmental and social responsibility groups have been pushing for more and more corporate performance information under the rubric of sustainability. Often these requests are for additional corporate transparency in areas that go beyond matters of sustainability. Furthermore, the requesters ask for information according to their own particular needs and formats, without much concern about the effect upon the reporting organization. The result is a great deal of confusion and frustration as companies are asked to respond to multiple information needs and frameworks.

The Global Reporting Initiative reporting framework

Several organizations have tried to address sustainable development reporting. The most notable is the United Nations-sponsored Global Reporting Initiative (GRI), which is creating a set of principles and a reporting framework that has broad applicability to both the reporting and user community. Since June 2000, the GRI has been issuing guidelines on sustainable development reporting, the stated purpose of which is to help organizations and their stakeholders communicate the reporting organizations' contributions to sustainable development. The GRI sees the emergence of "next-generation accounting," in which the new intangible assets—human and environmental capital, alliances and partnerships, brands and reputation—will be reported in much the same way as organizations now report tangible assets.[25]

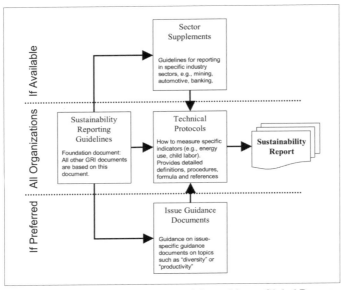

Figure 5-6: GRI family of documents. Adapted from: Global Reporting Initiative. http://www.globalreporting.org/guidelines/framework.asp.

In its most recent draft, the GRI guidelines contain over 100 indicators of sustainable development performance. The GRI also is beginning to publish technical protocols, detailed definitions, procedures, and techniques for measuring indicators. Finally, it is developing supplementary guidance for reporting by specific industry sectors. These documents fit into the GRI reporting framework depicted in figure 5-6.

Although targeted at corporations with substantial environmental and social impacts, sustainable development reporting may extend to engineering firms. According to the GRI guidelines, company sustainable performance extends beyond traditional boundaries of corporate responsibility, reaching out to its suppliers. Thus an A&E firm acting as a supplier of services is considered part of the value chain of the corporation, and its actions and policies relative to sustainability are counted as part of the overall impact of the corporation. Hence, clients that choose to report in accordance with the GRI guidelines may require their suppliers, including engineering firms, to provide data on their sustainable development performance. They may also require them to make commitments to sustainable development policies that are compatible with the policies of the corporation. Furthermore, sustainable development performance may extend beyond the operations of the A&E firm. It could include not only the use of energy and resources in their operations, but also how they apply sustainable development principles to the delivery of projects, such as the selection of materials, the disposal of construction wastes, and so on.

In addition to attracting the interest of environmental and social advocates, sustainable development reporting has become important to the financial community. An increasing number of investment groups view the sustainable development performance of corporations as a good indicator of long-term financial performance. There are now a number of sustainability indexes, including the Dow Jones Sustainability Index, FTSE4Good and Innovest's EcoValue '21, that track the performance of companies that have committed to sustainable development practices. In addition, there are a number of green funds starting to appear. Managers of these funds are seeking sustainable development performance information in order to create investment portfolios in companies that they believe will have better-than-market returns due, in large part, to their commitment to sustainability.

AA1000: a process for developing a sustainable development report

While the GRI guidelines provide guidance on what to report, the AA1000 series of assurance standards offers guidance on the reporting process. Created by the Institute of Social and Ethical AccountAbility, the guidance shows how companies can integrate their processes for stakeholder engagement to develop the appropriate sustainable development indicators, targets, and reports. The AA1000 guidance fills a gap between the reporting organizations

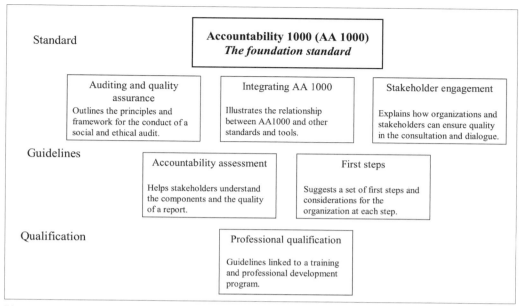

Figure 5-7: AA1000 Framework: AA1000 is an accountability standard for quality of sustainability reporting. Accountability includes three elements: (1) transparency to the stakeholders, (2) responsiveness in supporting continuous improvement, and (3) compliance with agreed upon standards. Source: AccountAbility 1000 (AA1000) Framework http://www.accountability.org.uk/upload-store/cms/docs/AA1000%20Framework%201999.pdf

and the users by providing a process that assures users that the information provided by the reporting organizations is credible, accurate, and balanced. The guidance offers procedures for a systematic "assessment of reports against three Assurance Principles."

- *Materiality: does the sustainability report provide an account covering all the areas of performance that stakeholders need in order to judge the organization's sustainability performance?*

- *Completeness: is the information complete and accurate enough to enable the reader to assess and understand the organization's performance in all these areas?*

- *Responsiveness: has the organization responded coherently and consistently to stakeholders' concerns and interests?[26]*

The AA1000 Assurance Standard was published in March 2003 and is available on the AccountAbility Web site.[27]

Greening your firm, that is, developing a set of programs to reduce your company's environmental footprint and improve its contribution to your communities, is a relatively small but important step. Not only does it offer proof of your interest and commitment to sustainability, but it also offers substantial cost savings in many facets of your operations.

---

[1] Guy Dauncey, Earthfuture—Stories from a Sustainable World (Gabriola Island, British Columbia, Canada: New Society Publishers, November 1999).

[2] Guy Dauncey, EarthFuture Consultancy, 395 Conway Road, Victoria, British Columbia. V9E 2B9, Canada, http://www.globalideasbank.org/socinv/SIC-84.HTML.

[3] See Earthday Network, Redefining Progress http://www.earthday.net/footprint/index.asp.

[4] See Redefining Progress: http://www.rprogress.org/programs/sustainability/ef/ef_household_0203.xls.

[5] Ecological Footprint Accounts (brochure from Redefining Progress, Oakland, CA).

[6] STMicroelectronics, Social and Environmental Report 2002 (Agrate Brianza, Italy May 2003).

[7] The 10 environmental commandments of STMicroelectronics, http://eu.st.com/stonline/company/environm/deca-log.htm.

[8] Business Week Online, "The Stars Of Europe – Managers, Pasquale Pistorio, Chief Executive, STMicroelectronics, "http://www.businessweek.com/magazine/content/02_24/b3787606.htm.

[9] Ibid.

[10] STMicroelectronics Company, "STMicroelectronics: Company Presentation," July 24, 2003, http://us.st.com/stonline/company/identity/slides/us.pdf.

[11] STMicroelectronics, "Environmental Decalogue," http://us.st.com/stonline/company/environm/decalog.htm.

[12] Earnest O. Robbins II, "Sustainable Development Policy."

[13] Earnest O. Robbins II, "Sustainable Development Policy."

[14] City of Austin, Capital Improvements Plan (Texas).

[15] The matrix can be found at http://www.ci.austin.tx.us/sustainable/matrix.htm

[16] W. Laurence Doxsey, "Case Study of Integrating Sustainability Evaluation into Public Building Projects" (paper presented at the Green Building Challenge, Vancouver, BC, 1998).

[17] Marie Logan, assistant to the School of Natural Resources and the Environment, University of Michigan.

[18] Alcoa, 2002 Sustainability Report, (Pittsburg, PA: Alcoa, 2003).

[19] These remarks were made by John Pizzey, executive vice president of Alcoa, at the annual meeting of the Aluminum Association in Nemacolin, Pennsylvania, on September 30, 2002.

[20] These ideas for greening were compiled from various sources including CH2M HILL, GreenBiz, and Global Environmental Management Initiative.

[21] Business for Social Responsibility's Web site can be found at http://www.bsr.org/index.cfm.

[22] The U.S. Business Council for Sustainable Development's Web site can be found at http://www.usbcsd.org/.

[23] Engineers Without Borders-USA's Web site can be found at http://www.ewb-usa.org/index.htm.

[24] GreenBiz.com, "Corporate Reporting," http://www.greenbiz.com/toolbox/essentials_third.cfm?LinkAdvID=25026.

[25] Sustainability Reporting Guidelines (Boston: Global Reporting Initiative, 2002), 3.

[26] Institute of Social and Ethical AccountAbility, "The AA1000 Assurance Standard," http://www.accountability.org.uk/aa1000/default.asp?pageid=52.

[27] Available at the Institute of Social and Ethical AccountAbility's Web site: http://www.accountability.org.uk/aa1000/default.asp.

CHAPTER 6

Delivering a Sustainable Project

Building sustainable buildings is a challenge for our contractors because it requires them to change the way they have been doing business.[1]

A school designed to "code" is the worst facility you can legally build.[2]

The challenge

Think about this scenario. Your client has a sizable and challenging project for you. Your assignment is to design and construct a rather large facility along with much of its associated infrastructure. The facility must not only be highly productive, but must achieve some level of sustainability, moving substantially beyond levels that can be achieved using conventional technologies. The client's organization has made a public commitment to sustainability and the client wants the facility to be designed and built in a way that truly contributes to sustainable development.

Figure 6-1: Use of an engineered wetland allowed the recycling of 7 million gallons per day of industrial wastewater. Photo courtesy of Jerry Jackson.

This hypothetical project contains a number of important design and operational questions concerning sustainable development projects:

- How can you design and construct this project so that it meets sustainability principles? How can you document and verify that the project contributes to sustainability?

- What is the best process for incorporating sustainable development principles into the delivery of a client project?

- Since sustainable development is a journey, how can we incorporate technological developments into projects in a way that advances the state of the practice, moving closer to conditions of sustainability?

- In the delivery of the project, what special considerations are necessary to incorporate successfully nonconventional (sustainable) technologies and systems?

Projects that truly contribute to sustainable development differ from projects designed using conventional technologies in several important ways. First, to be sustainable, a project must show improved performance (operational performance clearly above conventional) across one or more dimensions of sustainability. This requires a comprehensive set of sustainability goals and metrics against which the project can be designed and its performance measured.

Second, to show performance improvements, the sustainable project must be designed using achievements from other projects as benchmarks. Since sustainable development is a journey, progress can only be made by observing these achievements and then setting higher performance objectives to reach new levels of sustainability performance. This requires access to information about the achievements of others as well as information about new technologies that offer a promise of better performance. It is important to note that reaching new performance levels may not be possible on each and every project. However, projects that exceed conventional performance but do not exceed the sustainability benchmarks of others are very beneficial. They contribute by adding to the performance knowledge base, thereby helping to advance the standard of practice.

Third, projects that contribute to sustainability may incorporate newer and sometimes unproven technologies. Moreover, to achieve overall improved performance, it is more than likely that these technologies will interact as part of a larger system. This adds to the overall system complexity and increases control system requirements. Special care must be taken in the execution of the project to understand the interactions and to make sure the system is built to specification and adjusted to achieve maximum performance. To achieve project performance goals and objectives, some level of project commissioning is recommended.

Taken together, these differences necessitate different forms of delivery from those of conventional engineering projects. In conventional projects, the professionals generally act independently, each providing expertise from their particular discipline when called for. Such a delivery process tends to isolate the disciplines from one another, making it difficult to deal creatively with system interactions or new technologies. Furthermore, each discipline carries with it a set of assumptions, standards, procedures, systems, and technologies that the practitioners trust. Their approaches also come with a factor of safety and comfort. They follow approaches that they know will work under a wide variety of conditions and uncertainties, some of which are caused by the approaches and practices of other disciplines.

To achieve high levels of performance requires new forms of project delivery, forms that create a more collaborative atmosphere among the project team members. One approach that has gained attention is a process called a *design charrette*.[3] This is an intense and rigorous planning and design process conducted over a relatively short period of time. It involves a group of professionals from the project disciplines working together in a collaborative process to create a workable design. Openness and stakeholder involvement are essential ingredients. Cross-discipline teams working with stakeholders produce design solutions that are the product of multiple views and experiences. The compressed time schedule inspires creativity and discourages debate that is not on point. A well-run design charrette can bring about real change as the participants come to a new understanding of the problems and issues involved in the project.

Prerequisites for delivering a sustainable project

Several important conditions are necessary to achieve project success and to contribute to sustainable development:

- *A knowledgeable and committed project owner.* For the project to be successful, the project owner must be committed to making a substantive contribution to sustainable development. He or she must also be a knowledgeable buyer: knowing what is doable based on the achievements of others and knowing how the individual materials, equipment, and systems perform. The project owner should be supported by an executive champion, someone at a relatively high position in the organization who will reinforce when necessary the organization's commitment to sustainability.

- *A high-performance project team.* The project team assembled must not only contain the necessary mix of professional disciplines, but have the ability to collaborate to achieve project goals. The individuals on the team must be willing to put aside the conventional assumptions and practices of their discipline and work together to come up with improved solutions through an integrated approach to design. Mechanisms for stakeholder involvement and input are essential. Invited stakeholders should include not only public interest groups, but outside utilities, user groups, and other technical specialists.

- *Alternative procurement and contracting mechanisms.* Project team selection should be based on not only the team's aggregate experience and qualifications, but on its proven ability to work together to achieve improved performance over conventional performance. Performance-based contracting—providing incentive fees for performance over and above an agreed-upon base level—should be employed.

- *High but achievable sustainable development goals and objectives.* The project owner, working with the project team, must set goals and measurable objectives to achieve levels of performance that are substantively above conventional levels. These goals and objectives must be achievable based on the knowledge of what others have achieved and the performance of the requisite processes, systems, and technologies. With sufficient knowledge, the project owner may attempt to raise the bar on performance in one or more elements of sustainability. Finally, the project goals and objectives must relate to the whole-society goals of sustainability. That is, they must address the priority problems and issues of sustainable development recognized by society.

- *Access to and willingness to share knowledge and achievements.* Setting high but achievable sustainable development goals and objectives requires access to reliable information about the performance of the materials, equipment, systems, and technologies to be employed on the project. In order to make progress toward sustainability, practitioners must be able to learn from the experience of others and benchmark their projects against previous achievements.

Achieving sustainable project goals and objectives

Setting out sustainable project goals and objectives in one thing; achieving them is another. Project teams frequently encounter unexpected problems in meeting the levels of performance specified in the contract. Since the systems and technologies employed are relatively new and the team may be in uncharted territory, there is often a desire on the part of some team members or the team itself to compensate by lowering project goals and objectives. After all, they say, in this new arena of sustainable development, it may not be possible to achieve what we set out to achieve. Isn't achieving 80 percent of what we set out to do good enough?

The answer to this question should be no. If the team agreed that the project goals and objectives were achievable, then it is the duty of the team to work together and find the answers to problems as they arise. Bill Franzen, the executive director of operations for the Poudre School District in northern Colorado, has experienced this kind of team issue before. He observes that some teams, when faced with these sorts of project problems, tend to return to their old behaviors, blaming the newness of the technologies or the other project disciplines for failure to meet their commitments. His response as a project owner is to tell the team to solve the problem, referring them to their previous agreement on what was achievable. Franzen also recommends that project commissioning be used as early as possible in the project to make sure that the materials and equipment were installed as specified and that the systems are reaching expected levels of performance.[4]

Project delivery methodology

In this context, how do you deliver a sustainable development project successfully? Table 6-1 lists the key elements of project delivery and discusses the special considerations related to sustainable development.

Sustainable development project goals and metrics

As noted in the Preface, the journey toward sustainability will be accomplished incrementally, project by project. If we are to make efficient progress, then each project should be measured against a comprehensive set of sustainability criteria covering all aspects of sustainability. Projects that are intended to make a contribution to sustainability will do so only if performance improvements achieved in one or more dimensions of sustainability are not nullified by underperformance in other dimensions. In other words, when aggregated across all dimensions of sustainability, there must be a net gain.

Sustainable development project metrics and indicator systems are starting to appear in a number of places. In the United States, the U.S. Green Building Council has devised the Leadership in Energy and Environmental Design (LEED) certification system, a process for rating buildings in terms of their contributions toward sustainability. Using a somewhat different approach, the Netherlands-based firm the Arup Group Ltd., has devised a model called the Sustainability Project Appraisal Routine (SPeAR). This model provides a colorful display of sustainability project metrics together with a value judgment as to their relative rank. In addition, the International Federation of Consulting Engineers (FIDIC), an organization based in Geneva, Switzerland, has developed a framework and a process for setting comprehensive and meaningful project sustainability goals and measuring progress toward those goals. Embodied in FIDIC's Project Sustainability Management (PSM) guidelines, the framework is designed to ensure that a project's sustainability goals are aligned and traceable to recognized and accepted whole-society goals and priorities. The PSM process allows the customization of sustainable development project goals to suit local conditions and priorities.

Project complexity: no pain, no gain

Project complexity, both technical and nontechnical, is an issue in sustainable development projects. Designers seeking to incorporate sustainability principles into their projects and advance the state of the practice do so by adding features such as material recycling, resource conservation, and renewable energy sourcing into the design. Often these features require additional systems to be brought into the overall design and integrated with other systems. Furthermore, the incorporation of these added features likely require new technologies with which the engineers may have little experience.

Although the intent is to improve overall sustainable performance, additional systems and new technologies add complexity to the overall design and could decrease overall reliability and increase overall project design and construction costs. All of this suggests that managing a sustainable project requires a level of attention over and above a conventional project.

Delivering a sustainable project often presents additional complexity in the form of stakeholder involvement. By definition, projects that claim to contribute to sustainability cover a broad spectrum of issues beyond those of a conventional project. These groups will have new agendas and expectations, all of which need to be managed.

Achieving sustainable development project by project

If progress toward sustainability is to be achieved, it is essential to develop both a framework and a process for setting project sustainability goals and measuring progress. A framework needs to be constructed to ensure that a project's sustainability goals are aligned and traceable back to the whole-society goals and priorities established by the United Nations in Agenda 21 and in the Millennium Development Goals. In addition, a process is needed to guide the planning and delivery of projects. The process must accomplish the following:

- Assist the project owner and the engineer in developing practical project sustainable development goals, striking a balance between the project owner's aspirations and the issues of cost, achievability, and stakeholder concerns.

- Incorporate substantive stakeholder input throughout the project life cycle, ensuring that all substantive issues are addressed.

- Be open and transparent in terms of goals, stakeholder input, and project performance expectations.

- Provide mechanisms for feedback, assessment of results, sustainability performance benchmarking, and knowledge sharing.

An essential ingredient for this process is a comprehensive set of project sustainable development goals and indicators. These must cover the full range of sustainability issues and enable the practitioners to measure the specific project contributions toward sustainable development, all tying back to the whole-society goals of Agenda 21 and the Millennium Development Goals.

Table 6-1: Elements of project delivery and corresponding sustainable development considerations

I. Roles and responsibilities of the project manager		
Subelement	**All projects**	**Sustainable development**
Helping the client focus	Understand client needs and expectations.	Manage expectations. Clients may have a wide range of expectations regarding the implementation of sustainable development principles.
Creating a project vision	Create a vision of what the project will accomplish.	Work closely with client to create a sustainability vision. Develop high but achievable goals and objectives.
Building and maintaining the project team	Select the right people for the team; define assignments and responsibilities.	Create an integrated team in which all parties work together to achieve the project vision. Stakeholders are an integral part of the team.
Planning the project	Develop a work plan that meets the client's vision; plan for change.	Reach agreement on an achievable set of project sustainable development performance objectives and metrics.
Managing the resources	Manage time and money.	Recognize the need for schedule flexibility in order to test and verify technology performance. More time will be needed in the planning and design phases.
Ensuring quality	Meet or exceed expectations of the client and the stakeholders.	Meet or exceed the expectations of the client and the stakeholders. Work with the stakeholders in the early stages of the project.
II. Chartering, building, and sustaining the project team		
Chartering the team	Define the purpose, scope, goals, behaviors, roles, and responsibilities.	Place emphasis on working together to achieve overall project sustainability goals and objectives.
Building the team	Improve skills: problem-solving, interpersonal, and decision-making skills.	Emphasize team integration. Foster a desire and ability to work through technology and system integration issues.
Sustaining the team	Optimize technical and behavioral aspects of the team.	Push for achieving the "stretch" goals of sustainability. Emphasize that achieving, say, 80% of the agreed-to goals is not good enough.

Table 6-1: Elements of project delivery and corresponding sustainable development considerations

Subelement	All projects	Sustainable development
III. Developing the work plan		
Defining the project	Create project objectives, scope, and the work-breakdown structure.	Factor in the requirements for systems integration. Incorporate commissioning.
Determining project resources, schedule, and budget	Select project personnel; define the schedule; budget costs.	Select people who are committed to achieving the goals of the project.
Defining project instructions	Define client involvement, procedures for public involvement, and how the project will be controlled.	Provide for additional consideration for stakeholder involvement, public scrutiny.
Determining quality management	Determine how to deliver what the client wants on time and on budget.	Include project commissioning to make sure that all equipment and systems were installed as specified and the specified operating performance has been achieved.
Developing a client service plan	Develop a communication plan between client and consultant.	Create a plan that helps manage expectations and change.
Developing a change management plan	Devise a decision framework for managing change.	Benchmark what others have achieved and compare to client's goals. Client and consultant may need to reset expectations if goals cannot be reached.
Developing a closure plan	Develop procedures for closing the project.	Record performance and compare with goals and objectives. Project commissioning is especially important.
IV. Endorsing the project		
Positioning for endorsement	Describe the benefits of endorsement, get client and stakeholder acceptance.	Obtain project team buy-in as well as stakeholder buy-in. Expect a diverse set of stakeholders.
Signing off	Formally accept the project charter and work plan.	Make sure the charter and the plan are accepted by all stakeholders.
Endorsing change	Have all stakeholders make a commitment to accept change.	Expect change when applying new and relatively untried technologies. However, this should not be an excuse for missing project goals and objectives.
V. Managing change		
Identifying the change	Develop process for identifying change; determine type, source, and impact.	Identify potential for changes early in the project.
Analyzing the effects	Verify change, impacts, and value.	Expect interactions among systems. Changes in one area can affect others and result in missed project objectives. Effects may be harder to quantify.
Developing a response strategy	Ask, what needs to be done and when? Cost? Time to do?	Avoid the temptation to accept 80% of design performance as good enough. Seek to meet design goals and objectives.
Communicating strategy and gaining endorsement	Communicate the change and necessary response to the client and get endorsement.	Involve the team in determining the solutions. Involve stakeholders.
Revising the work plan and monitoring effects	Ensure the work plan is revised to reflect changes; monitor the effects of change.	Monitor and document changes for future reference for sustainability benchmarks.

Table 6-1: Elements of project delivery and corresponding sustainable development considerations

Subelement	All projects	Sustainable development
VI. Closing the project		
Phasing in the closure of tasks	Close individual tasks as they are completed.	Recognize that cost creep can occur if documentation is not done systematically.
Demobilizing	Systematically demobilize staff from the project as tasks are completed.	Expect unanticipated requirements during systems integration.
Recommending staff development	Recommend further staff training based on lessons learned.	Provide opportunities for staff to learn the application and integration of new systems and technologies.
Performing technical closure	Summarize results of technology application; transfer technology.	Document technology application and performance in detail.

A critical component—an environment for innovation

To achieve conditions of sustainability, a prerequisite for success is the establishment of an *environment for innovation*: working conditions in which learning and creativity are fostered and celebrated. In this environment, project owners, observing the achievements of others, are encouraged to set stretch goals, seeking to establish new and higher benchmarks for sustainability performance. At the same time, engineers are free to try out new approaches, test new technologies, and replace old ways with new and more sustainable alternatives.

Openness and transparency are the essential ingredients of this environment. Project owners and engineers must engage stakeholders in dialogue throughout the project development, design, and delivery processes to ensure that their issues and concerns are fully considered. This stakeholder engagement process operates in two directions. Stakeholders must voice their issues and concerns about the project so that the project owner and the engineer can incorporate those matters throughout the project life cycle. On the other hand, the project owner and the engineer must educate the stakeholders as to the current state of the practice as well as the limitations of what is currently achievable. To make this an efficient and effective process, these parties must establish an atmosphere of trust and collaboration.

The speed of progress, as always, will be governed by the ability of engineers to innovate: to imagine, invent, develop, test, and apply new processes, systems, and technologies to the problems at hand. However, in the case of sustainability, problem definition is elusive, driven as much by public perception as by technological fact. In these changing conditions, progress will be marked by a series of fits and starts, triggered by events and politics as well as investments and accomplishments. Here, the role of the engineer is key, contributing logic and structure in a climate of uncertainty and confusion.

The role of project sustainable development goals, objectives, and indicators

While project goals and objectives set the direction, project sustainable development indicators provide the means by which progress can be measured. The purpose these indicators is to enable project owners, engineers, and stakeholders to gauge real progress toward sustainability by comparing actual sustainable performance achieved on the project against sustainability goals, the intended project performance. To these ends, a comprehensive set of project sustainable development indicators is an essential tool for measuring actual accomplishments, demonstrating transparency to the stakeholders and building a knowledge base for the practitioners.

To function properly, this indicator set must be grounded in the overarching principles, goals, and priorities of sustainable development. In addition, it must be sufficiently comprehensive to cover all relevant aspects of sustainability, yet be of a size that is manageable and effective for communication. Also, the indicator set must allow for customization in order to align to local requirements and conditions. Finally, the process by which these accommodations are made must be open and transparent.

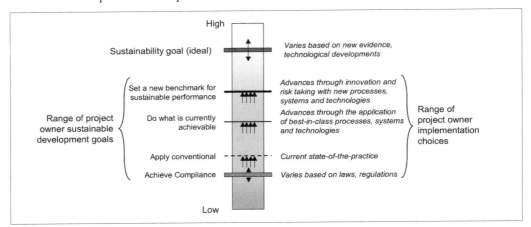

Figure 6-2: Sustainable Development objectives and indicators conceptual model. Source: International Federation of Consulting Engineers, Sustainable Development Task Force, Project Sustainability Management Guidelines.

A conceptual model of a sustainable development project indicator is presented in figure 6-2. In this model, the range of sustainable performance is characterized on a generic scale from high to low. Conditions of sustainability are achieved somewhere in the high range, but that threshold is variable based on local conditions, evidence about resources and carrying capacity, and any technological developments that could alter the definition of sustainability for this particular indicator. For example, new knowledge of ecological carrying capacity limitations could shift the sustainability objective to higher levels. In contrast, the invention of a low-cost, energy-efficient desalinization technology could alter dramatically the availability of freshwater and thus drop the corresponding sustainability targets.

In setting sustainable development project goals and objectives, the project owner has a number of choices. At the low end of the sustainability scale, he or she may do nothing more than apply conventional technology, directing the engineer to implement the current state of the practice. The *state of the practice* is defined as the current procedures and technologies normally applied by engineering professionals in the field. In the model, it is assumed that on the depicted sustainability scale, the current state of the practice lies just above some compliance level defined by local or national laws or regulations or by global treaties. In many, if not most cases, there may not be any laws, regulations, or treaties associated with sustainable development project indicators.

Alternatively, the project owner may decide to make a contribution to sustainable development by applying processes, systems, or technologies that hold promise to perform substantially above conventional approaches. In this case, the project owner and the engineer may assess what others have accomplished on similar projects. At the same time, they may also identify new processes, systems, and technologies that hold promise for setting new levels of performance. Once this assessment is completed, the project owner and the engineer can set goals for sustainable performance in one or more aspects of sustainability, as measured by the corresponding indicators. Here, they may decide to match what others have accomplished on other projects using previously applied approaches.

Under the right conditions, the project owner may attempt to set new levels of performance using new and relatively untried approaches. As the project progresses, sustainable project performance indicators will enable the engineer to measure and record performance.

Overall, each success in the application of more-sustainable processes, systems, and technologies will raise incrementally the definition of "best in class." As they are used repeatedly on projects, applications once considered advanced eventually will be reclassified as state of the practice, driving up the definition of conventional. Over time and given the right environment, the range of implementation choices will move toward the high end of the spectrum, eventually reaching conditions of sustainability.

Relationship of sustainability indicators to whole-society goals

The principles, goals, and priorities for sustainable development across our entire society are well documented in Agenda 21. Following the Earth Summit, the United Nations empowered its Commission on Sustainable Development (CSD) to devise a set of sustainable development indicators that measure and calibrate progress toward sustainable development goals, all based on the issues, goals, and priorities identified in Agenda 21. Accordingly, the CSD created a list of sustainable development indicators organized inside a framework of sustainability themes and subthemes, all traceable to the whole-society issues, goals, and priorities of Agenda 21. The CSD's indicators were intended to translate whole-society goals into a form accessible to decision makers at the national level to be used as guidance in crucial decision making. They are considered to be important tools to communicate ideas, thoughts and values to help countries make informed decisions about sustainable development.

If progress toward sustainable development is to be made through individual projects, then the CSD's whole-society indicators must be translated to project-level indicators. Furthermore, these project-level measures must be comprehensive, containing all of the key components of sustainability. Omission of any of these components will distort the evaluation and call into question the project's value and contribution. As an example, a project with the objective of, say, reducing the amount of water used might accomplish this by additional energy use or by increasing the use of toxic materials. Failure to include all of the elements of sustainability may achieve one sustainability goal at the expense of others, resulting in minimal or negative net progress.

Lastly, project sustainability indicators serve as guideposts and benchmarks. This helps ensure that progress in one aspect of sustainability is not made at the expense of others. Importantly, they show what others have achieved, inspiring everyone to set new and higher levels of sustainable performance.

Other sustainability indicator systems

Based on the outcomes of the Rio Summit and the publication of Agenda 21, numerous sets of sustainable development indicators continue to be produced by different organizations for a variety of purposes. Some of these indicator sets, classified by their intended purpose, are shown in table 6-2.

Organizations including the United Nations, standard-setting institutions, multinational unions, national and regional governments, local authorities, financial organizations, and public interest groups have understood that if society is to embark on a course toward sustainability, new measures and criteria on which to gauge current status and progress will be required. Many have devised various sets of sustainability indicators reflecting their own needs and perceptions. Some of these sets are intended to measure whole-society conditions

Table 6-2: Sustainable development indicator classifications

Name	Description	Examples
Whole-society indicators: sustainability of a particular geography or political boundary		
Global	Overall assessment of the current state of the world, mapped to Agenda 21	UN CSD, Pilot Analysis of Global Ecosystems, Millennium Ecosystem Assessment, Ecological Footprint
Regional and local	Response to Local Agenda 21; assessment of sustainability factors determined to be important to the local population	Pastille, Sustainable Seattle, Santa Monica, NRTEE
Organization-based indicators: sustainability of the operations of an organization		
Industry, government, or nongovernmental organizations	Indicators of how an organization is performing in terms of a set of sustainability indicators	Global Reporting Initiative
Investor-based indicators: correlation of corporate sustainability with financial performance		
Project risk assessment	Principles, processes, and indicators for managing project risk	The Equator Principles
Financial performance indices	Any published index that tracks the financial performance of companies that have committed to sustainable development practices	Dow Jones Sustainability Index, FTSE4Good, Innovest: EcoValue, 21
Green funds	Funds that hold investment portfolios in companies that believed to have better-than-market returns because of their commitment to sustainability	Domini Social Equity Fund, Triodos Bank
Project-based indicators: assessment of a project's contribution to sustainability		
Project screening	Indicators for screening projects as to their likelihood of achieving sustainability results	World Bank, the Equator Principles
Project performance	The actual contribution a project makes toward sustainable development; includes efforts made in the construction phase	FIDIC's Project Sustainability Management, SPeAR, CRISP, BEQUEST, CH2M HILL's four-step project screening process, LEED

of sustainability. Others are used as investment tools in which corporate sustainable development commitment and performance are seen as leading indicators of future financial performance. Still others are used to measure corporate performance against their organization's own interpretation of sustainable development. Finally, some groups have created sustainable development indicators for projects in the built environment using qualitative scoring methods to rate projects, highlight areas of exceptional performance, and identify areas for improvement. All of these indicator approaches have an appropriate place and application. However, they do not explicitly and fully connect projects back to the fundamental issues, goals, and priorities of sustainable development as defined at the Rio Summit.

Project Sustainability Management (PSM)— the FIDIC approach

In undertaking the task of translating whole-society sustainability indicators to project level indicators, FIDIC has recognized the following[6]:

- *Sustainable development is a whole-society concept.* Any attempt in measuring a project's contribution to sustainability must be based on complete and generally accepted principles of sustainable development.

- *Sustainable development is a moving target.* Current concepts of sustainable development problems and issues will be altered substantially by the course of events and as new knowledge develops. Furthermore, it is likely that these changes will occur within the life cycle of typical engineering projects. Thus, the indicators used at the start of a project will likely be very different than those used at the deconstruction and decommission stage.

- *The issues and impacts related to sustainable development are often location and culture dependent.* Many sustainable development issues and impacts that are significant in one part of the world may be unimportant in another. Others such as climate change, ozone depletion, and deforestation are ubiquitous. In many parts of the developing world, freshwater, sanitation, health, and jobs are critical issues.

- *A prerequisite for making progress toward sustainable development is the establishment of an environment for innovation.* Progress can only be made if the practitioners have the freedom to explore, invent, test, apply, and evaluate promising processes, systems, and technologies that offer better and more sustainable performance. This requires a high degree of openness and transparency in order to foster understanding among the stakeholders and knowledge development and sharing among the practitioners.

FIDIC's Project Sustainability Management guidelines describe how project owners and engineers can incorporate the principles of sustainable development into individual projects. The components of the system are twofold:

1. A framework of sustainable development goals and the corresponding indicators, both of which map back to the whole-society issues, goals, and priorities of Agenda 21, and the corresponding sustainability indicators developed by the United Nations Commission on Sustainable Development

2. A process for setting and amending sustainable development project goals and indicators, making them consistent with the vision and goals of the project owner, compliant with Agenda 21, and tailored to local issues, priorities, and stakeholder concerns

FIDIC has developed a set of core sustainable development project goals and indicators organized in a framework that corresponds to the whole-society issues, goals, and priorities of Agenda 21. Also, FIDIC has devised a process to amend these goals and indicators, allowing them to be customized to actual project conditions while retaining its whole-society scope. In addition, the process addresses the full life cycle of the project, from concept development through design, construction, operation, deconstruction, and disposal. In this sense, project sustainability goals and indicators become part of the overall project delivery process.

The conceptual model of the PSM sustainable development project goals and indicators framework is presented in figure 6-3. Sustainable development issues are divided into categories: environmental, economic, and social. Below each category, sustainability issues are organized into themes and subthemes. Each subtheme is associated with one or more indicators of sustainable development. As noted earlier in figure 6-2, each indicator can be characterized by a spectrum of sustainable performance achieved in relation to the current state of the practice, applicable laws or regulations, and sustainability goals.

The approach offered by the PSM differs from approaches that others have taken in the following ways. First, it enables the user to customize the project indicator set based on its relevance to the project scope, conditions, and context while maintaining a close connection to the whole-society issues, goals, and priorities of sustainable development. Second, the process takes into account the changing capabilities of processes, systems, and technologies. In the journey toward sustainable development, the invention and successful application of

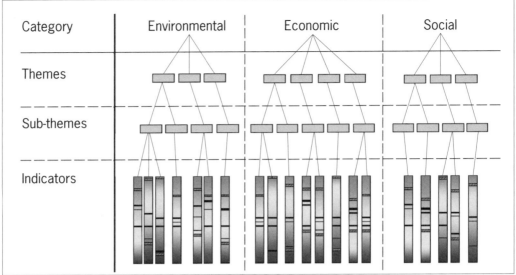

Figure 6-3: Conceptual model of the PSM project goals and indicators framework. Source: International Federation of Consulting Engineers, Sustainable Development Task Force, Project Sustainability Management Guidelines.

new processes, systems, and technologies will advance the state of the practice, setting new benchmarks for sustainable performance. As the practice advances, what was once best in class will eventually become conventional, replaced by new and better approaches. Third, this process provides a mechanism for establishing performance benchmarks, comparing the achievements of others and setting ever-higher goals for sustainable performance. Even though whole-society sustainability issues are well established, the approach to these issues and the available technology for dealing with them is evolving very rapidly.

The advantage of PSM over other indicator systems

For a project to make a valid and verifiable contribution to sustainable development, it must be designed and delivered in a way that produces a measurable net positive impact across all the dimensions of sustainability—environmental, economic, and social—throughout the project life cycle. This is not an easy task. In the absence of any overall guidance, governments, NGOs, public interest groups, and others have produced a substantial array of sustainability measurement systems based on their own particular interests and agendas. Indicator systems based on such a narrow focus are difficult to relate to the balancing of alternatives that takes place in real projects, and may even create conflicting targets. In contrast, PSM starts with a broad set of goals and indicators, grounded in generally accepted whole-society principles of sustainability. Using the process, project owners and their engineers can modify them to reflect local conditions as well as the range of potential solutions.

The value of PSM to a project owner is substantial. The starting point, the core set of goals and indicators, is virtually unassailable because it is founded in the original concepts of sustainability. Moreover, it recognizes the realities of sustainable development: the fact that progress will be incremental and perfection elusive. It focuses on achieving incremental improvements based on the accumulation of knowledge and experience and through innovation. Finally, it allows the project owner to demonstrate his/her organization's contribution toward sustainability in a way that is both transparent and verifiable.

The value of PSM to the engineering industry is also substantial. To effectively respond to the challenge of sustainability and to meet project owner needs, the engineer gains a deep understanding of the project owner's objectives and performance over the project's entire project life cycle. This creates a close client relationship and an innovative environment that enables the consultant to deliver services that are clearly based on quality and special knowledge of sustainable practices and technologies. PSM also adds a new dimension of project management that covers processes related to indicator development and application and operates in parallel with the established areas of cost, time, scope, human resources, risk, procurement, communications, and quality management. It expands the engineer's scope of services, enabling them to add sustainable development to its portfolio of client service offerings.

The PSM process

Project Sustainability Management offers a process for establishing, demonstrating, and verifying a project's contribution to sustainable development. If a project owner wishes to incorporate sustainability goals into his or her project, PSM provides a process by which those goals can be credibly established in concert with accepted whole-society goals and priorities. Progress toward those goals can be measured and verified against measurable and agreed upon indicators.

The process is designed to be highly transparent in order to create and maintain stakeholder trust. It is recognized that progress toward sustainable development will only happen if project owners, engineers, and stakeholders work together, creating and applying new and more sustainable processes, systems, and technologies.

PSM addresses a broad range of project sustainability management issues:

- How to integrate project owner sustainability goals into a project
- How to show the connection between sustainability achievements of a specific project and recognize whole-society sustainable development goals and priorities
- How to create and maintain transparency in the development of sustainability goals and the corresponding project sustainability indicators
- How to incorporate the goals and needs of a wide range of stakeholders
- How project sustainability goals and indicators can affect project objectives and designs

The core set of project indicators

The starting point for PSM is a set of "core" sustainable development indicators, derived from national and international sustainability goals and targets. As such they are tied to the original whole-society objectives of Agenda 21, adjusted to make them relevant to projects.

Indicators are observed or calculated parameters that show the presence or state of a condition or trend. They are the tools for measuring and monitoring progress toward goals, providing a basis for judging the extent to which progress has been made or corrective action is required. They are also an important management tool for communicating ideas, thoughts, and values. As the United Nations Commission on Sustainable Development observed, "We measure what we value, and we value what we measure."[7]

In a typical project management system, indicators are used to measure important project performance parameters. Measurements are compared against expected results, and corrective actions are initiated if indicator measurements are sufficiently outside expected values. For PSM, the emphasis is not so much on corrective actions, but on measuring performance and contributing to the sustainability knowledge base. The primary value of project sustainability indicators comes from measuring and learning what the application achieved rather than how it was controlled.

Indicators are typically built in a two-part process. The first part is a framework that ties the indicators to overall objectives defined in the global context. Frameworks also help in stimulating proposals about what should be measured—categorizing issues, organizing ideas, and establishing a common vocabulary.

The second part of the indicator building process involves the development of a series of conceptual models that describe the performance of the project in each framework category in terms of measurable parameters. These parameters describe the cause and effect relationships within the context of the project and ensure that the selected indicators can be used to measure project performance. Indicators are selected for a particular project if they are seen to influence outcomes and respond to changed external factors.

As an example, the framework for a global objective of improved health in our society might include the subcategory of drinking water, with a conceptual model that if safe drinking water were available to a higher percentage of the population, global health might improve. The indicator would then become the percentage of the population having access to safe drinking water. Access to safe drinking water in terms of a project in the developed world might involve issues of maintaining water quality coming out of the taps in each building. In the context of the developing world, the level of improvement might be much more rudimentary and involve a central, safe drinking water source for the community. In a project-specific way, it would be clearly advantageous to engage the local affected community in a dialogue about the feasibility and applicability of the systems and approaches to be used in the delivery of safe drinking water and in the indicators that might be used to describe the results of the project.

The recommended approach for developing indicators is therefore to use this two-step process. First, the framework is used to map overall sustainability goals to specific issues that will be addressed. Second, a conceptual model is used to develop indicators that map to the issues. This process ensures that indicators map back to global objectives but are project specific in detail.

Working on the mandate from the 1992 Earth Summit and Agenda 21, the CSD converted the conditions and priorities of sustainable development spelled out in Agenda 21 into sustainability indicators. The CSD used a theme framework organized around four major sustainable development categories: social, environmental, economic, and institutional. Each major category was broken into themes and subthemes that covered the significant elements of sustainability and mapped to the chapters of Agenda 21. This framework was then used by the CSD as a basis for identifying and selecting the corresponding indicators of sustainability. Using this process, the CSD developed 65 indicators covering 15 themes and 38 subthemes.

The PSM core indicators were derived from the CSD framework. In the FIDIC PSM model, a set of 42 indicators (14 themes and 30 subthemes) was derived from the 65 CSD indicators, based on their relevance to projects. The Institutional category was not included in the group of PSM categories, as the themes and subthemes contained therein were seen as not relevant to projects.

Each indicator has been adjusted to make it relevant to project activities. For example, the CSD indicator "Percent of Population Living below Poverty Line" was converted into an appropriate PSM project indicator, "Proportion of workers or companies employed on the project from the local area." In this example, it is assumed that the project owner and the engineer cannot affect substantially the percentage of the population living below the poverty line. However, they can contribute to poverty reduction in the region by employing local workers.

Table 6-3: FIDIC PSM core sustainable development project indicators [8]

Theme	Subtheme	PSM core project indicators
Social		
Equity	Poverty (3)	SO-1: Proportion of workers or companies employed on the project from the local area
Equity	Gender equality (24)	SO-2: Existence of hiring and wage policies related to minorities and women employees
		SO-3: Proportion of minorities, women hires
		SO-4: Wage comparison of minorities, women compared with standards
Health (6)	Sanitation	SO-5: Proportion of population with access to adequate sewage treatment
Health (6)	Drinking water	SO-6: Proportion of population with access to safe drinking water
Health (6)	Healthcare delivery	SO-7: Proportion of population with access to primary healthcare facilities
Health (6)	Occupational safety and health	SO-8: Record of safety performance during construction
Human rights	Child labor	SO-9: Record of the use of labor during project construction
Housing (7)	Living conditions	SO-10: Proportion of persons living with adequate floor area per person
Population (5)	Population change	SO-11: Change in number and proportion of populations in formal and informal settlements affected by the project
Culture	Cultural heritage	SO-12: Assessment of impacts on local culture, historic buildings
Culture	Involuntary resettlement	SO-13: Degree to which the project displaces the local population
Integrity	Bribery and corruption	SO-14: Efforts to monitor and report bribery and corruption
Environmental		
Atmosphere (9)	Climate change	EN-1: Quantities of greenhouse gases (GHGs) emitted in all phases of project
Atmosphere (9)	Ozone layer depletion	EN-2: Quantities of ozone-depleting substances used in all phases of project
Atmosphere (9)	Air quality	EN-3: Quantities of key air pollutants emitted in all phases of project
Atmosphere (9)	Indoor air quality	EN-4: Quantities of indoor air pollutants
Land (10)	Agriculture (14)	EN-5: Proportion of arable and permanent crop land affected by this project
Land (10)	Agriculture (14)	EN-6: Quantities of fertilizers used compared with norms
Land (10)	Agriculture (14)	EN-7: Quantities of pesticides used compared with norms
Land (10)	Forests (11)	EN-8: Extent to which forests are used or affected in the development, design, and delivery of the project
Land (10)	Forests (11)	EN-9: Extent to which wood is used in all project phases
Land (10)	Desertification (12)	EN-10: Extent to which land covered by project is affected by desertification

Table 6-3: FIDIC PSM core sustainable development project indicators (continued) [8]

Theme	Subtheme	PSM core project indicators
Environmental (continued)		
Oceans, seas, and coasts (17)	Coastal zone	EN-11: Measurements of changes in algae concentrations
Oceans, seas, and coasts (17)	Coastal zone	EN-12: Changes in populations living in coastal areas
Freshwater (18)	Water quantity	EN-13: Measurements of water usage on project during all phases
Freshwater (18)	Water quality	EN-14: Measurements of biological oxygen demand (BOD) on water bodies affected by project during all phases
Freshwater (18)	Water quality	EN-15: Measurements of fecal coliform in freshwater bodies affected by project during all phases
Biodiversity (15)	Ecosystem	EN-16: Proportion of area affected by the project that contains key ecosystems
Biodiversity (15)	Species	EN-17: Measurements of affect of project on the abundance of key species
Economic		
Economic structure (2)	Economic performance	EC-1: Extent to which project provides economic benefit to local economy
Consumption and production patterns (4)	Material consumption	EC-2: Extent of use of materials compared with norms, other practices
Consumption and production patterns (4)	Energy use	EC-3: Extent of energy consumption compared with norms, other practices
Consumption and production patterns (4)	Energy use	EC-4: Extent of use of renewable energy resources compared with norms, other practices
Consumption and production patterns (4)	Waste generation and management (19–22)	EC-5: Quantities of industrial and municipal wastes generated compared with norms, other practices
Consumption and production patterns (4)	Waste generation and management (19–22)	EC-6: Disposition of industrial and municipal wastes compared with norms, other practices
Consumption and production patterns (4)	Waste generation and management (19–22)	EC-7: Quantities of hazardous wastes generated compared with norms, other practices
Consumption and production patterns (4)	Waste generation and management (19–22)	EC-8: Disposition of hazardous wastes compared with norms, other practices
Consumption and production patterns (4)	Waste generation and management (19–22)	EC-9: Quantities of radioactive wastes generated compared with norms, other practices
Consumption and production patterns (4)	Waste generation and management (19–22)	EC-10: Disposition of radioactive wastes compared with norms, other practices
Consumption and production patterns (4)	Waste generation and management (19-22)	EC-11: Extent to which waste recycling and reuse is employed in all phases of project, compared with norms, other practices
Consumption and production patterns (4)	Transportation	EC-12: Measurements of transportation modes and distances traveled by people and materials in all project phases; comparison with norms, other practices
Consumption and production patterns (4)	Durability (service life)	EC-13: Extent to which durable materials were specified; design for extended service life
Consumption and production patterns (4)	Care, ease of maintenance and repair	EC-14: Extent to which the facility requires care and maintenance, compared with norms

In the PSM process, the list of core indicators can be modified and expanded in a series of steps, applying local sustainable development indicators and stakeholder input. The process does not, however, allow an indicator to be deleted or materially modified if it is seen as germane to project sustainability.

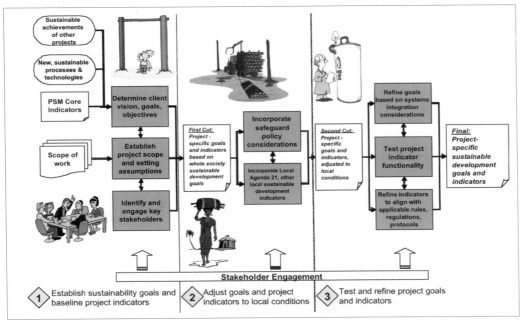

Figure 6-4: The Project Sustainability Management (PSM) process. Source: International Federation of Consulting Engineers, Sustainable Development Task Force, *Project Sustainability Management Guidelines*.

Creating project-specific sustainable development goals and indicators

The PSM process for creating and implementing sustainable development goals and project indicators is depicted in figure 6-4 and summarized in the following:

- *Stage one.* Establish sustainability goals and baseline project indicators:
  - o Determine the project owner's vision, goals, and objectives for the project.
  - o Establish the project scope and setting assumptions.
  - o Identify and engage key stakeholders.
- *Stage two.* Adjust goals and project indicators to local conditions:
  - o Incorporate applicable safeguard policy considerations.
  - o Identify and incorporate Local Agenda 21 or other local indicator development activities.
- *Stage three.* Test and refine project goals and indicators.
  - o Refine goals based on systems integration considerations.
  - o Test project indicator functionality.
  - o Refine indicators to align with applicable regulations and protocols.

A description of each of the stages is provided in the following sections.

Stage one: Establish sustainability goals and baseline project indicators

Stage one covers the early planning and design phase of the project. Here, the project owner and the engineer work together to incorporate sustainable development goals into the project scope and align them to the project owner's overall vision for the project. Key stakeholders are identified and engaged at this point. In addition, the project owner and the engineer may review what others have achieved in terms of sustainable development goals and what new processes, systems, and technologies have emerged.

Establish the project scope and setting assumptions

The first step in the process is to obtain a detailed understanding of the project: scope, setting, intended use, and so on. The purpose of this step is to understand the nature of the project and its potential environmental, economic, and social impacts. It is assumed that the project owner has conducted a project risk analysis and has determined that there are no known or anticipated conditions or circumstances that could stop the project.

In this step, the project manager uses the PSM core sustainable development themes and indicators as a checklist to make sure that the project is sufficiently well defined in order to assess all aspects of project sustainability. Some of these concern the physical aspects of the project, for example, the constructed works, including expected usable life operations, maintenance requirements, and disposition at the end of the facility life. Others refer to the material and energy flows. Still others refer to the behavior of the user. A partial list of these information requirements is given in the following:

- Detailed project scope
- System boundaries, both physical and temporal (life cycle)
- Affected environments, habitats, or other environmentally sensitive areas
- Affected groups (affected both positively and negatively, such as involuntary displacement or resettlement, new employment)
- Affected cultural heritage sites
- Life expectancy of the constructed works
- Requirements for service, maintenance, and refurbishment
- Demolition, deconstruction, recovery, recycling, and disposal
- Energy consumption and related impacts such as greenhouse gas emissions
- Material usage
- Energy changes and evolution of energy chains
- Expected usage of the facility
- Access to transportation[9]

Determine the project owner's vision, goals, and objectives for the project

What does the project owner want to accomplish in terms of sustainable development? To answer this question, the engineer should work with the project owner to determine the environmental, economic, and social goals for the project. The project owner may have specific sustainability goals in mind such as saving water, providing a diversity of mobility options, or employing alternative energy resources.

In this step, the project owner's vision, goals, and objectives for project sustainability are compared against the PSM core sustainable development project goals and indicators to make sure that the elements of sustainability the owner intends to pursue can be matched. It is important to recognize that project planning and design decisions made at this stage will have impacts throughout the project life cycle, that is, through construction, operation, and deconstruction and disposal.

Table 6-4: Example impacts of planning and design decisions on sustainability

Project life cycle phase	Decision	Impacts
Development	Location, function, partnerships, financing, cost	Access, quality, occupant efficiency, occupant comfort, contribution to the community
Planning, siting, and design	Use of recycled materials, openness of design, use of natural lighting, access to transportation	Material intensity, energy efficiency, occupant efficiency, occupant comfort
Construction	Disposition of construction wastes, recycling	Construction environmental footprint, use of recycled materials
Operation	Energy efficiency, indoor air quality, materials usage	Occupant efficiency and productivity
Deconstruction, disposal	Ability to recycle building materials, building reuse	Resale value, redevelopment potential

Additional guidance on opportunities for innovation are found later in this chapter.

Identification and engagement of key stakeholders

In the past, companies recognized a limited number of groups—shareholders, employees, regulatory agencies, the financial community, and a few others—as legitimate stakeholders. Today the situation has changed considerably. Enabled by information technology and telecommunications, many nongovernmental organizations and activist groups are emerging with new powers of information acquisition and communication. If they see fit, these organizations can use the power of the media and the Internet to communicate what they perceive as misbehavior on the part of any organization. In effect, these groups set de facto standards for governmental and nongovernmental behavior. They can have a strong impact on the organization's reputation and a correspondingly strong (positive or negative) impact on its financial performance.

It is the same situation with projects. Organizational performance of the project owner is judged in part by the projects he or she designs, constructs, and operates, and thus can be the subject of inquiry for stakeholder groups. Therefore, it is important that the owner and the engineer identify early the key stakeholders for the project, understand their issues and information needs, and establish a set of project indicators that meet those needs. It is a recognized source of sustainability conflict when the interests of local stakeholders run contrary to the perceived interests of society as a whole.

Stage two: adjust goals and project indicators to local conditions

In stage two of the PSM process, goals and indicators established at stage one are modified to reflect local conditions, particularly the conditions and concerns of the low- and middle-income nations.

Incorporate applicable safeguard policy considerations

If the project is located in a low- or middle-income country as defined by the World Bank Development Indicators database,[10] then additional indicators should be developed to reflect the special sustainability concerns and policy safeguards applicable to developing nations. These concerns include natural habitats, pest management, forestry, dam safety, indigenous peoples, involuntary resettlement, cultural property, child and forced labor, and international waterways.[11]

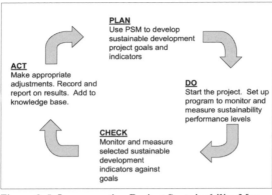

PLAN
Use PSM to develop sustainable development project goals and indicators

ACT
Make appropriate adjustments. Record and report on results. Add to knowledge base.

DO
Start the project. Set up program to monitor and measure sustainability performance levels

CHECK
Monitor and measure selected sustainable development indicators against goals

Figure 6- 5: Incorporating Project Sustainability Management into a project quality management system. Source: International Federation of Consulting Engineers, Sustainable Development Task Force, Project Sustainability Management Guidelines.

In addition to safeguard policies, additional consideration should be given to the local resources and capacity to understand and apply policies pertaining to sustainable development. Without this understanding, certain categories of issues may dominate the discussions with stakeholders, perhaps to the exclusion of all others, no matter what their importance to sustainability happens to be. For example, in South Africa many, if not most, project owners and engineers have not been exposed to the concepts of sustainable development. Currently, social and socioeconomic issues are receiving considerable attention while environmental issues are not. Therefore, in working in the low- and middle-income countries, practitioners should be careful to understand the context in which PSM is applied, making sure that local priorities are being addressed, but within the context of the sustainable development core project indicators.

Identify and incorporate Local Agenda 21 or other local indicator development activities

The locality in which the project will be delivered may have developed its own set of whole-society indicators in accordance with chapter 40 of Agenda 21. If that is the case, then the engineer should locate the indicator set and use it to develop project-specific indicators. Here, the engineer should compare the local sustainable development indicators with the PSM core indicators. The engineer should add any local indicators that are not present in the core indicator set or modify the core indicators to obtain better agreement with local indicators.[12] Even if no Local Agenda 21 processes are under way, the engineer still should check with local government officials and environmental groups to identify other sustainable development indicator-related activities.

The outcome of this step is a set of sustainable development indicators based on the PSM core sustainable development indicators, but modified to make them relevant to local problems and conditions. The indicators must have a substantial level of endorsement from local government officials or key stakeholder groups.

Stage three: test and refine project goals and indicators

Moving into stage three, the project owner and the engineer have established a set of sustainable development project goals and a corresponding set of sustainable development project indicators, reflecting the owner's vision of the project and modified to reflect local conditions and sustainable development goals. In this final stage, the owner and the engineer make three additional refinements to these goals and indicators.

Refine goals based on systems integration considerations

In the design and delivery of a project that advances the state of the practice across several dimensions of sustainable development, the engineer draws upon a number of processes, systems, and technologies to achieve the desired results. At this stage, it is important for the designer to consider how these individual elements will work together to achieve the desired outcome. Here, these elements are cross-checked to make sure that the interferences are minimized. It may be necessary to change the sustainable development goals established earlier to accommodate these integration considerations.

Test project indicator functionality

Once the indicator list is complete, the project owner and the engineer should review the list of indicators and test them to see if the process has produced a sensible and workable indicator set. The Pastille Consortium suggests a three-part test that it claims will help the practitioner by stimulating different ways of thinking, creating better understanding of the context in which the indicators will operate, and, ultimately, identifying indicator strengths and weaknesses.

The following list outlines the Pastille test.[13] Additional details may be found in the referenced guide.

- *Map the indicator profile.* At what level are the project indicators in use: strategic, program, or project? Within which tools are the indicators operating? What is the indicator typology? What is the purpose of the indicator system? Who are the stakeholders relevant to the project's indicator system? What is their role? Are they supportive, neutral, or obstructive? How can they become more supportive of the project's indicators?

- *Assess the arena of action.* This part is intended to help the practitioner define the shape of the arena for action and better understand the factors, both positive and negative, that affect the use of the indicators. These include issues of stakeholder identification, engagement, communication, trust, and cooperation; ease of data collection; and indicator linkage to targets and thresholds and how indicators are used to make decisions.

- *Picture the arena of action.* Interpret and present the results of the analysis. The authors suggest plotting the results on a radar chart as a way of comparing each dimension of the evaluation.

Applying these tests may turn up problems with the indicators chosen. The engineer and the project owner should conduct a final review of the indicator set and make any necessary modifications.

Refine indicators to align with applicable rules, regulations, and protocols

For some projects, client organizations, associations, local authorities, or other institutions may require the application of existing project indicator sets that relate to sustainability. Furthermore, for reasons of overall image and reputation management, competitiveness, or as part of the overall enterprise strategy of the organization, the project owner may want to apply other indicator protocols to the project. Some examples are presented in the following:

- *Reporting in accordance with the Global Reporting Initiative (GRI) Sustainability Reporting Guidelines.* Some client organizations have made it a policy to report their sustainable development performance in conformance with the GRI guidelines.

- *Achieving LEED certification.* The project owner may want to achieve a certain certification level for one or more of the buildings or facilities in the project. In that case, the project owner may need to add or modify some of the project indicators to match the LEED system. For example, the U.S. Department of Defense has issued a directive stating that its new facilities will meet a certain level of LEED certification.

- *Evaluating the project using SPeAR.* The project owner may want to add or modify the project indicators to match the SPeAR appraisal model.

If needed, the indicators developed under the PSM process may be adjusted to align with the required appraisal or reporting protocols.

Final sustainable development goals and project indicators

Practitioners who complete these stages of the PSM process will have developed a set of sustainable development goals and the corresponding indicators for their project.

Sustainable Project Appraisal Routine (SPeAR)

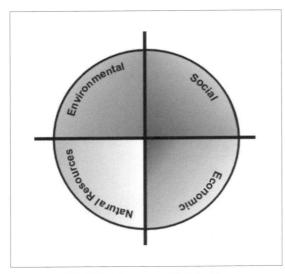

Figure 6-6: SPeAR® 4-Quadrant Model with core sectors. Source: Arup Group Ltd. Reprinted by permission.

The Sustainable Project Appraisal Routine is based on a four-quadrant model that structures the issues of sustainability into a framework that can be used either as a management information tool or as part of a design process. Created by the Arup Group, Ltd., SPeAR brings sustainability into the decision-making process by focusing on four elements: environmental protection, social equity, economic viability, and efficient use of natural resources.

According to Arup, the information generated by the appraisal prompts innovative thinking and informed decision making at all stages of design and development. This allows continual improvement in sustainability performance and assists in delivering sustainable objectives.

Similar to John Elkington's shear zones discussed in chapter 2, the SPeAR model recognizes the linkages between environmental, social, economic, and natural resource and systems. These systems cannot function in isolation, and the intersections between environmental and social systems, between social and economic systems, and between environmental and economic systems determine quality of life, socioeconomic conditions, and eco-efficiency respectively. Sustainability lies in the integration of all these systems.

The appraisal is based on the performance of each indicator against a scale of best and worst cases. Each indicator scenario is aggregated into the relevant sector, and the average performance of each sector is then transferred onto the SPeAR diagram. The logical and transparent methodology behind the SPeAR diagram ensures that all scoring decisions are fully audit traceable; the dia-

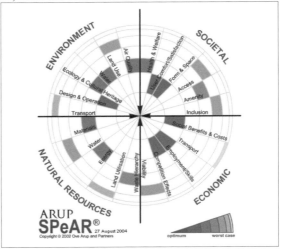

Figure 6-7: SPeAR® 4-Quadrant Model with hypothetical scoring. Source: Arup Group Ltd. Reprinted by permission.

gram provides a unique profile of the sustainability performance, highlighting both strengths and weaknesses from the perspective of sustainability. The sectors of SPeAR and the underlying indicators are not weighted, as this would assume that some sectors make a greater contribution to sustainability than do others.

The appraisal is undertaken at a particular time; however, sustainability is not static and, as sustainability drivers change, so will the potential for further gains in environmental protection, social equity, economic viability, and resource efficiency.

LEED Green Building Rating System

The LEED Green Building Rating System is a process for applying and scoring buildings based on a set of project indicators. Developed and administered by the U.S. Green Building Council (USGBC), it offers a framework and procedures for comprehensively evaluating and scoring buildings based on sustainable development principles. Buildings are scored on a point scale ranging from zero to 69 and can achieve four levels of certification.

- LEED certified: 26–32 points
- Silver level: 33–38 points
- Gold level: 39–51 points
- Platinum level: 51+ points (out of 69 possible points)

LEED was created by the USGBC to standardize the definition of a green building and to promote the design of buildings that are environmentally benign, while being more productive—a contradiction in terms as seen by many developers, architects, and engineers. According to the USGBC, commercial and residential buildings account for over 65 percent of U.S. electricity consumption, 36 percent of greenhouse gas emissions, and 12 percent of potable water usage. Construction and demolition of buildings produces 136 million tons of waste annually. At the global level, buildings account for 40 percent of the raw materials used—about 3 billion tons per year.[14] Thus, designing and constructing buildings that use less energy and raw materials, and produce less waste, can contribute substantially to sustainable development.

Although criticized as offering an oversimplified and incomplete assessment of contributions to sustainability, LEED appears to be transforming the building market. First, it is raising stakeholder awareness of the benefits of owning, operating, and working in a green building by introducing life cycle analysis and integrated building design into the purchasing decision. In the past, building purchasing decisions were based primarily on the first cost, not the full cost, of operation. Now, using LEED, owners can see the benefit of investing in design and construction phases in order to reduce operating costs, the largest cost of ownership.

Second, the LEED rating system and its four levels of certification have stimulated competition to improve building performance. This competition involves not only architects, builders, and owners, but also companies seeking to develop and market high-performance building technologies.

Third, LEED certification tends to increase the value of buildings by raising the market awareness of not only the environmental, but also the economic and performance benefits of green buildings. Buildings that are more productive and cost less to operate are worth more money to the owner.

Fourth, LEED certification helps to prevent "greenwashing." By setting standards coupled to a trustworthy certification program, uncertified claims of "greenness" are seen as not credible and are discounted.

LEED has been gaining rapidly in popularity across the United States. Agencies in the federal government, including the General Services Administration, the Department of Defense (U.S. Army, Navy, and Air Force), the Department of Energy, and the U.S. Environmental Protection Agency have endorsed LEED. The U.S. Army Corps of Engineers has devised its own version of LEED, called SPiRiT. This model takes into account the specific needs of the Department of Defense installations. Some of these agencies have gone a step further by incorporating the achievement of LEED certification into building and facilities procurement. Six states and 10 cities now use LEED certification in new building construction. Today the USGBC has 62 certified and over 850 registered building projects.

CH2M HILL's sustainable project road map

Several years ago, the sustainable development practice group at CH2M HILL asked these questions: How do you deliver a project that is demonstrably more sustainable than what could be achieved using conventional technologies? How far can you reasonably go in extending sustainable performance beyond that which conventional technologies can achieve? How do you make those design decisions?

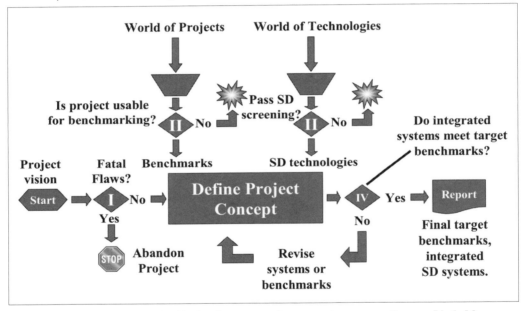

Figure 6-8: CH2M HILL's sustainable development project screening process. Source: Linda Morse, CH2M HILL, Presentation to the American Council of Engineering Companies, Environmental Business Action Coalition Sustainable Development Meeting, January 2000.

In response, sustainable development architect Paul Bierman-Lytle and a CH2M HILL design team developed a framework and methodology for incorporating sustainable development principles into facilities and infrastructure projects[15]. The methodology consists of a series of five evaluation steps, culminating in a design report used to guide the detailed design and delivery. The approach is to draw upon new and more-sustainable technologies and evaluate them against sustainability design criteria, known technology performance, systems integration, economics, and client goals. This five-step process is depicted in figure 6-8.

The process was developed for a proposed refurbishment of a hotel, school, dormitory, and residences on a 1,900-acre site in Southern California. One of the project's prime objectives was to incorporate sustainable development principles to the extent practical.

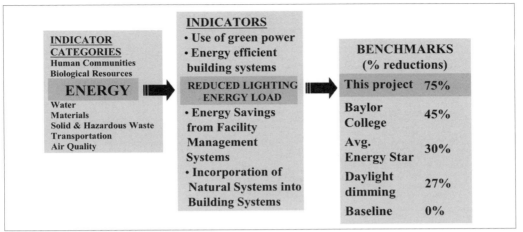

Figure 6-9: Benchmarking and "raising the bar." Source: Linda Morse, CH2M HILL, Presentation to the American Council of Engineering Companies, Environmental Business Action Coalition Sustainable Development Meeting, January 2000.

The process is based on five basic questions:

1. *What do we want to accomplish in regard to sustainability?* What is to be accomplished by way of moving in the direction of sustainable development? What are our project's vision and objectives?

2. *Are there any fatal flaws?* Is there any issue about location, design, or operation of the proposed project that would cause the project to be stopped?

3. *What levels of sustainability performance have been achieved by others?* What has been achieved on other projects and what can be applied to this project?

4. *Do the technologies and systems we are proposing meet sustainable development criteria?* How do these stack up against sustainability principles?

5. *Will the technologies and systems work together and meet sustainability cost and performance objectives?* Are the technologies and systems compatible? Can we meet our cost performance goals or will we have to reset them?

Step 1: Project vision

The first step is to decide on a project vision. In this step, the client and the engineer work together to develop the vision of what sustainability objectives the client wants to accomplish with this project. Some examples of project sustainability objectives offered by Linda Morse follow:

- *Water—Water consumption will be substantially less than for traditional comparable developments.*

- *Energy—Significantly reduce energy consumption, with respect to comparable development facilities, using green building design, facility management, demand reduction, efficient building systems, and other methods.*

- *Biological Resources—In areas where natural communities will be modified, enhance and maintain ecosystem features using best management practices.*

- *Human Communities—Design the project to promote a sense of place and to provide opportunities for frequent, casual interaction of residents, visitors, and employees.*[16]

These overarching objectives provide a guide to the specific designs and technologies employed.

Step 2: Fatal flaw analysis

In the second step, we consider any serious or fatal flaws in the project that could prevent the project from going forward. Some examples are cultural resources, geology, air quality, and traffic.

Step 3: Benchmarking

The purpose of step 3 is to learn what others have achieved in the way of improvements in technologies, processes, equipment, or techniques that resulted in progress toward sustainability. Here, the engineers use some system of sustainable development project indicators to track performance. These indicators were derived from a set of indicator categories that cover all aspects of sustainability.

The Natural Steps System Conditions	Examples of Specific Sub-criteria
1. Substances from the earth's crust can not systematically increase in nature	Does the action use or release naturally occurring materials from earth's crust that are toxic to health of plants and animals?
2. Substances produced by society can not systematically increase in nature	Does the action cause emissions to the air of persistent man-made substances?
3. The physical basis for the productivity and diversity of nature must not be systematically diminished	Does this action affect the ability of land to replenish surface and ground water?
4A. Fairness in meeting basic human needs	What is the extent to which an approach supplies diverse living environments?
4B. Efficiency in meeting basic human needs	What is the degree to which energy use is minimized?

Figure 6-10: The Natural Step System Conditions and criteria examples. Source: Linda Morse, CH2M HILL, American Development Corporation, Paul Bierman-Lytle, personal conversations, 2003.

Using these categories and indicators, engineers look for projects in which new technologies were applied and had achieved some level of improved performance over conventional technologies. They also define baseline performance—performance achieved by conventional technologies. Last, based on the client's objectives for the project, the engineers may be asked by the client to raise the bar on performance by setting a goal at some new and higher performance level. A sample application of an indicator for water use is shown in figure 6-11.

Step 4: Technology screening

In addition to the technologies already applied on existing projects, it is essential to identify and screen new technologies. As the interest in sustainable technologies rises, engineers should expect to see an array of new technologies that may offer better performance. The intent of the process is to screen these technologies, as well as the technologies identified in step 3, using the principles of sustainability. CH2M HILL chose to use the system conditions of the Natural Step as a basis for screening, and developed the corresponding evaluation criteria. Figure 6-10 provides examples of how specific subcriteria might be developed using the system conditions of the Natural Step.

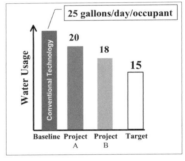

Figure 6-11: Raising the Bar on water conservation. Source: Linda Morse, CH2M HILL, Paul Bierman-Lytle, personal conversations, 2003.

Step 5: Systems integration

The last and perhaps the most important step in the process is the integration of all the systems specified in the project. At this step the question is asked, do the systems and technology function together cohesively? While a technology may be able to achieve a high level of performance on its own, that may not be the case when it is integrated with other project systems. It is also at this step that the costs to build, operate, and maintain the project are considered.

The process approach is to bring together a multidisciplinary project team to evaluate the systems, consider cross-system impacts, and seek to confirm that the systems will operate as planned and without impacting negatively another area. For example, if a selected water treatment technology was determined to have high-energy usage, this would conflict with an energy-use reduction objective. Through each iteration, the team evaluates the indicator goals and may reset them depending on how conflicts in objectives are resolved. Hopefully, the net result will be a project that is a clear improvement over conventional technologies.

Projects that incorporate the principles of sustainability and seek to raise the bar across one or more sustainability parameters generally involve a number of systems that must be integrated.

Eco-efficiency as a source of innovation

Organizations are finding that the principles of eco-efficiency are a useful starting point for embarking on the journey toward sustainable development. These principles offer new perspectives, ways of rethinking how products are made or how services are delivered. Table 6-5 offers a long list of ideas for innovation based on the seven principles of eco-efficiency.

Table 6-5: Eco-efficiency checklist [17]

Reduce the material intensity of goods and services
• Can the consumption of water be reduced? • Would the use of higher-quality materials create less waste at the later stages of the project? • Can wastes be reused on-site or transported elsewhere for reuse? • Can products or services be combined to reduce overall materials intensity? • Can packaging be reduced or eliminated?
Reduce the energy intensity of goods and services
• Can we find ways to employ renewable energy? • Would the use of different materials reduce energy usage? • Can we use the waste heat from one process to supply another? • Can building energy usage be monitored and controlled? • Can transportation be reduced or made more efficient?
Reduce toxics dispersion
• Can toxic substances be totally eliminated from the process or product? • Can wastes and emissions be reduced during the project construction phase? • Are there ways to better handle harmful materials during the construction phase? • Can we specify products with low toxic or harmful substance content?
Enhance material recyclability
• Can the wastes be remanufactured, reused, or recycled? • Can we specify products with high-recycled content? • Can we design the facility or infrastructure giving consideration to reuse, flexibility, or recycling? • Can we design the facility or infrastructure giving consideration to ease of deconstruction and material recovery?
Maximize the sustainable use of renewable resources
• Can we specify products made with resources certified as sustainable? • Can we design facilities and infrastructure, making maximum use of passive heating and cooling? • Can we employ renewable energy sources in our design?
Extend product durability
• Can we incorporate durability considerations in our designs? • Can we design the facility and infrastructure for ease of maintenance? • Can we design in a high degree of flexibility?
Increase the service intensity of products
• Can we work with clients to improve the service intensity of their business model, i.e., find ways to sell more service associated with the product instead of selling the product itself? • Can we add more knowledge content to the services you sell? • Can we leverage your knowledge of the client's business to reduce the client's costs or reduce waste? • Can we expand our scope of services to meet increased stakeholder needs?

Major companies have turned to eco-efficiency as a screening tool in product development. For example, BASF uses eco-efficiency analysis as a strategic tool for making products more economical and environmentally friendly. It has created proprietary software called Ecologistics to compare data on economics, environmental impact, and responsible care to find the best solutions for its customers and products. Using this tool, economic and ecological advantages and disadvantages of BASF's products and processes are compared directly against one another. Working with the University of Karlsruhe and the Öko-Institut, the tool is now being expanded to a three-dimensional graph that compares environmental effects, social effects, and costs.[18] Project teams at Procter and Gamble use the Product Sustainability Assessment Tool (PSAT) to determine whether a project is consistent with sustainable development principles. Through a series of questions, the tool generates a sustainability profile graphic, which the team then uses to strengthen sustainability characteristics and shore up weaknesses.

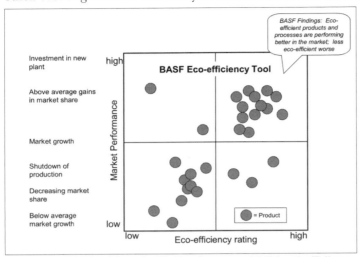

Figure 6-12: BASF and Sustainable Development - Walking the Talk, Presentation by Barry John Stickings at the Financial Times Conference "Are We Minding The Gap", London, May 28, 2002. http://www.basf.de/en/corporate/sustainability/publikationen/?id=CjmT.4ZJlbsf0I1

Tunneling through the cost barrier

Many of the innovations catalyzed by sustainable development involve whole-systems engineering, that is, optimizing not just parts of the system, but the system as a whole. Amory Lovins's organization, the Rocky Mountain Institute (RMI), has been studying this issue and notes that the conventional wisdom for optimizing energy efficiency is too narrowly focused, often missing huge opportunities for saving energy and money.

Using the conventional approach, organizations make energy-saving investments by analyzing each investment in isolation from others, stopping investments when the costs outweigh the benefits. However, RMI staffers point out that this sort of isolated thinking ignores the interactions among other parts of the system and may cause analysts to miss huge benefits. In an article in the summer 1997 issue of the *RMI Newsletter*,[19] the authors point to a home air-conditioning example. By analyzing components and subsystems separately, they were able to cut air-conditioning requirements by 66 percent. However, by paying the extra costs associated with the energy savings that heretofore were rejected because they could not save enough energy to pay for themselves, the designers were able to eliminate the air-conditioning system entirely, bringing in substantial savings in overall capital costs.

RMI identifies four principles for tunneling through the cost barrier.

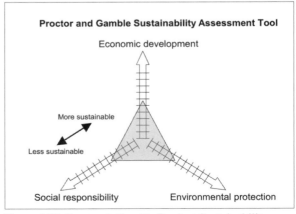

Figure 6-13: Proctor & Gamble Product Sustainability Assessment Tool (PSAT). Source: Proctor & Gamble.

- *Capture multiple benefits from a single expenditure.* Practitioners should think about what additional benefits may be derived from optimizing on a single variable. In the example cited above, the designers had reduced energy costs and found that, by extending beyond the normal cost-benefit barrier, they were able to reduce overall capital costs.

- *Start downstream to turn compounding losses into savings.* In chapter 1, I noted a Rocky Mountain Institute example of how the use of larger pipes in industrial piping design can save considerable capital and operating costs. This was the work of Jan Schillam, an engineer who designed an industrial pipe and pumping system by starting downstream looking at the pipes first. By using larger-diameter pipes, straight piping layouts, and more insulation, Shillam was able to reduce substantially the size of the pumps and motors, saving on construction, energy, and capital costs. He also increased reliability and lowered operation and maintenance costs.

- *Get the sequence right.* Lovins offers a process for achieving big energy savings by addressing the design task in the right sequence:[20]

  o *People before hardware.* Look at the service you intend to deliver to people and see if, for example, by changing the location or configuration, you can reduce the need for the hardware (facilities or equipment) in the first place.

  o *Shell before contents.* Get the design of the building or facility right to avoid having to "fix" or adjust the contents to make up for faulty design.

  o *Application before equipment.* Think about what service you are trying to provide and see what changes could be made that would reduce the amount of equipment needed to accomplish it.

  o *Quality before quantity.* Can you achieve the same result by improving the quality of the service? In the case of lighting, seek to reduce glare first, not overcome glare by adding more light.

  o *Passive before active.* Think about how to use natural systems first, for example, sunlight, heating, and cooling, and equipment second.

  o *Load reduction before supply.* Think about ways to reduce the demand for services before designing for supply.

- *Optimize the whole system, not the parts.* Understand the workings of the whole system to identify the interactions of the various components and what other costs might be affected.

A common goal for engineering firms and their trade associations has been to make a substantial shift toward the value pricing of engineering services. This goal is based on the belief that engineering services are very much undervalued and ought to be priced somewhere near the value the services bring to the client, as opposed to rates for professional hours. While this goal seems sensible enough, the engineering community has found that goal extremely difficult to reach. In this highly fragmented industry, every firm that tries to price its services on value is quickly supplanted by a host of firms that do not.

This situation is made worse by an industry that has trouble demonstrating good value to its clients. In project procurements today, almost every firm has the requisite technical skills to not only make the cut, but also make its technical qualifications indistinguishable from its competitors. As a result, the selection comes down to price, particularly in a tight economy.

How do you change the basis of competition? Earlier in this book, I noted that sustainable development is one of those issues that happens once or twice a decade, one that has the potential to change the way business is run. Today organizations in industry and government are engaged in a major makeover. They are reducing their environmental impacts and forging new relationships to their stakeholders. They see sustainable development as a powerful new market force and they are responding with new products and services.

Today the engineering community sits in an enviable position. The movement toward sustainable development will take place through a succession of engineering projects designed to make operations and infrastructure more sustainable. These projects will require the development and implementation of new and more sustainable technologies. This is not only the work of engineers from many disciplines, but it is new work requiring new knowledge and skills, few of which reside in any firm. Firms that can acquire this unique knowledge and these skills will find that they not only differentiate themselves technically, but also can start to price their services based on the value provided.

Attached to this opportunity is a serious threat. Other firms outside the engineering community have not only seen this opportunity, but are actively delivering sustainable development-related services and are taking market share. As I noted previously, accounting firms have seen the emergence of sustainable development reporting among their clients. They have responded by extending the financial auditing practice into environmental and social auditing and turning up new projects to fill perceived performance gaps. Architects have not only started to design so-called green buildings, but they have set the standard for what it means for a building to be declared green. I classify these examples as disturbing—a slow response to a confusing market. However, what is frightening to me are the examples of engineering shortsightedness exposed by Amory Lovins, Paul Hawken, Eng Lock Lee, William McDonough, Michael Braungart, and others. By neglecting energy costs, full-cost accounting, and other aspects of whole-systems thinking, we diminish the value of the profession. Unless we change, we deserve to be priced out as a commodity service.

How shall we change? First and foremost, the engineering profession needs to value innovation—not just talk about it, but also embed it into its practices. We ought to establish research and development functions within our organizations and spend some time devising new services and adapting new technologies to better meet client needs. Right now, we wait for our clients to tell us what to do. While research and development organizations may be out of reach for most firms, they can develop partnerships with universities, collaborating to convert academic research into working applications. Having an eye on the marketplace, we can guide university research to better match future needs. Through these efforts, we can perhaps develop enough foresight to take better advantage of new issues as they appear.

---

[1] Stu Reeve, energy manager, Poudre School District. Personal conversation. July 2004.

[2] From a slide presentation, Pathways to Creating Sustainable Schools, given by George Brelig, AIA, RB+B Architects, Inc. and Michael Spearnak, AIA, Poudre School District. Oct. 2003.

[3] The term charrette comes from the French word meaning "cart." Originally, it was used to describe the intense, last-minute efforts by architecture students studying at the École des Beaux Arts in Paris in the 1800s. Students' project drawings were collected on a cart or charrette circulated by the proctors. Architects now use the term to describe intense, short-term design efforts.

[4] Bill Franzen, in discussion with the author, July 2004.

[5] Project Delivery: A System and Process for Benchmark Performance (Denver, CO: CH2M HILL, 1996).

[6] FIDIC's Project Sustainability Management system was developed by the FIDIC Sustainable Development Task Force. This task force is chaired by the author, who developed the system and drafted the guidelines. The full set of guidelines together with project examples will be published by FIDIC in fall 2004. See www.fidic.org for details.

[7] Indicators of Sustainable Development: Framework and Methodologies. Background Paper No. 3. United Nations Commission on Sustainable Development, Ninth Session, New York, 16-27 April 2001. p. 2.

[8] Source: International Federation of Consulting Engineers, Sustainable Development Task Force, Project Sustainability Management guidelines. The numbers in parentheses denote the chapter in Agenda 21 to which the theme or subtheme refers.

[9] Adapted from Tomonari Yashiro, project leader, Building Construction/Sustainability in Building Construction, ISO document no. SO/TC 59/SC 17 N 35 (Geneva Switzerland, International Organization for Standardization, September 18, 2003), 6.

[10] The World Bank Development Indicators database can be accessed at the following location on the Internet: http://www.worldbank.org/data/countryclass/classgroups.htm.

[11] Guidance on policy safeguards can be found at the following Internet location: http://www.equator-principles.com/exhibit2.shtml.

[12] A survey of Local Agenda 21 processes can be found at the following Internet Web site: http://www.iclei.org/rioplusten/final_document.pdf.

[13] Pastille Consortium, Indicators in Action: A Practitioners Guide for Improving Their Use at the Local Level (London, UK, Pastille Consortium, European Union, May 2002), 22–30.

[14] U.S. Green Building Council, An Introduction to the U.S. Green Building Council and the LEED Green Building Rating System ([Washington, DC:, U.S. Green Building Council] August 2003).

[15] Linda Morse, "A Sustainability Roadmap: Creating the Sustainable Development Project" (paper presented at the American Council of Engineering Companies, Environmental Business Action Coalition's Winter 2001 Conference: Business Profiles in Sustainable Development, Marco Island, FL, February 2001).

[16] Ibid., 4.

[17] Source: Livio DeSimone and Frank Popoff, Eco-efficiency (Cambridge, MA: The MIT Press, 1997), 85–88.

[18] Barry Stickings, "BASF and Sustainable Development: Walking the Talk" (presentation to the FT-International Finance Corporate Conference, London, UK, May 28, 2002).

[19] "Tunneling Through the Cost Barrier," Rocky Mountain Institute Newsletter 13, no. 2 (Summer 1997): 1–4.

[20] Ibid.

CHAPTER 7

The Future of Sustainable Development

The person who does not worry about the future will shortly have worries about the present.[1]

A changing marketplace

This book has argued that sustainable development is becoming a major influence on business and government, creating new opportunities for the engineering firms that can discern the trends in their particular markets and position themselves for the ensuing projects. In the last 15 years, the sustainable development movement has shifted from a somewhat depressing United Nations prediction to a concept that has inspired industry and government leaders to rethink the basic purpose and goals of their organizations. Since the Brundtland Commission report was issued in the 1987, hundreds of industry and government organizations have adopted the principles of sustainability. More important, they are changing over their operations, finding ways to do more with less. These organizations are using less energy and materials to produce the same or better quality of goods and services, and they are making money. At the same time, they are reducing their environmental footprint and accepting more societal responsibilities. In turn, these changes have created new requirements for scientific, engineering, and design talent that is tuned in to the needs of sustainable development.

For many in the engineering community, this shift in the marketplace has been elusive. Over a thousand companies, many of which are clients of engineering firms, have joined the World Business Council for Sustainable Development (WBCSD) or one of its regional councils. Only a handful of engineering firms can be found in its rosters. In these organizations, the prerequisite for membership is a public commitment to sustainability. Many of these same companies have joined the Global Environmental Management Initiative, an organization that assists its members in achieving environmental and social performance excellence, sharing knowledge, experience, and tools among its members. Others participate in an organization called Business for Social Responsibility, which provides tools, education, and advice to companies, helping them incorporate socially responsible business practices into their operations. The Global Reporting Initiative, the institution set up by the United Nations to write sustainable development reporting guidelines, reports that 340 organizations now reference the guidelines in its sustainable development reports. Another organization, the Corporate Social Responsibility Newswire Service,[2] provides over 300 company reports on its Web site. Each of these reports contains not only the company's achievements, but also areas needing improvement.

Some of the engineering firms are catching on. Companies like CH2M HILL, Hatch Corporation, Kennedy/Jenks Consultants, Montgomery Watson Harza, and Nolte Associates have active sustainable development practices. Over the last five years, members of the International Federation of Consulting Engineers (FIDIC) have become recognized by the United Nations, the International Chamber of Commerce, and other international organizations as the primary engineering voice for sustainable development. At the World Summit on Sustainable Development, FIDIC, along with the WBCSD and the International Chamber of Commerce, presented the engineering industry's contribution to sustainable development.[3]

Late in 2002, while many in the U.S. engineering trade organizations were still debating policies on sustainability, over 4,000 architects and engineers attended an overbooked U.S. Green Building Council convention in Austin, Texas, looking primarily at the organization's

Leadership in Energy and Environmental Design (LEED) certification program.[4] One of the attendees reported back that he had not seen this many people at a conference so stirred up over an issue since the hazardous waste cleanup programs of the 1980s.[5]

Still, in the midst of all this interest, the sustainability movement has a long way to go. Sustainable development and all of its issues and benefits are still substantially unknown or ignored. Even with proof that building green can save money, not only over the life of the building but in the up-front design and construction costs (first costs) as well, many developers and contractors are not bothering to consider sustainable designs or obtain LEED certification.[6] In the absence of some major defining event, it appears that the sustainability movement will progress slowly, responding to client and public interest group pressures rather than to rational design improvements.

Purpose of this chapter

Thus, for engineering firms thinking about adding sustainability to their portfolio of practices, a number of important questions need to be answered. How soon will the concepts of sustainability reach the firm's client base? What form will sustainability take? What kinds of engineering services will be needed? What actions might the engineering firms (or the engineering community) take in order to shape the issue? What should the engineering firms do to stay at the forefront of this issue, at least in their own domain?

The sustainability movement is in a relative state of infancy, which suggests that there is still plenty of opportunity to shape the movement both in scope and direction. In this chapter, we will look at the trends and market drivers that will likely shape the sustainability movement over the next five, 10, or 20 years. Because of the movement's infancy, much of the discussion will concentrate on the near term, adding particular emphasis on opportunities for project identification, positioning, and competitive differentiation.

Thinking about the future

It is not possible to make accurate predictions about future trends and events. However, it is possible to make systematic analyses of trends and patterns from which can be drawn reasonable expectations of how the world might unfold. Based on this kind of knowledge, however ambiguous, we can plan a course of action, identify the uncertainties, and proceed according to that plan, making changes as needed. Recognizing what we know as well as what we don't know enables us to better plan for and react to changes when they occur.

Today the world is changing faster than perhaps we once thought possible. The events of September 11, 2001, produced a sea change in the way we now think about threats to our security. New technologies such as computers, telecommunications, genetic engineering, the Internet, nanotechnology, and composite structures are producing a vast array of economic and social changes that our societies are finding difficult to cope with. The sustainable development movement is also ushering in changes in the way we think about resources, the environment, and social responsibility. All of these things will cause, or are causing now, significant upheaval in both our businesses and our daily lives.

Learning as much as we can about possible future trends, events, and issues is now a fundamental necessity. Firms that can identify and track these matters can position themselves to respond to opportunities created by changing client needs and stakeholder concerns. Likewise, they can alertly identify and respond to potential threats.

Tools for studying the future

The most significant problem facing everyone who attempts to study the future is how to sift effectively though the myriad of information sources and pull out those trends worthy of further study and tracking. Today in this fast-changing environment those of us who track trends are in a constant battle to prevent information overload as we try to screen, assess, and rank trends that may coalesce into issues important to our business or our clients' businesses.

Futurists, those people that study the future, use a number of techniques to think about and sketch out future possibilities:

- *Trends analysis.* By collecting and analyzing selected indicators of local, regional, or world conditions, for example, population growth, air or water pollution, or per capita consumption, one can develop forecasts of future conditions through simple extrapolation or by developing more-complex mathematical models.

- *Precursor analysis.* By looking for patterns in the movement of certain indicators, the practitioner might be able to distinguish stages of change that could then be applied to predicting future changes.

- *Scenario analysis.* By creating of a series of possible futures in narrative description form that cover the range of plausible futures, the practitioner can consider how certain trends and events need to unfold (or not unfold) for a scenario to be realized. Scenario analysis is a useful tool for thinking about trends with high plausible impacts but for which the probability of occurrence is not known.[7]

The strategic planning literature offers a number of approaches for systematically identifying, screening, and processing trends, events, and issues for further study and prioritization. For this book, I have distilled the process into five steps:

1. *Set up a trends screening matrix.* Prepare a matrix of important stakeholders versus trends. The matrix should be set up in the form of a worksheet as shown in figure 7-1. For convenience and ease of assessment, I have sorted the trends into categories: economic, technological, political-legal, and sociocultural. However, readers may wish to modify these or use other categories.

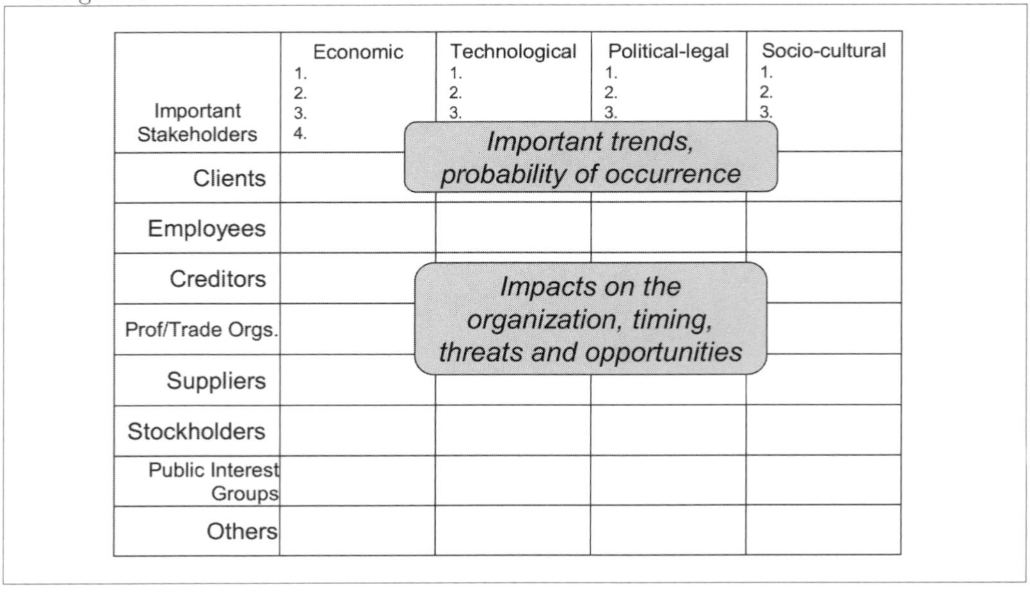

Important Stakeholders	Economic 1. 2. 3. 4.	Technological 1. 2. 3.	Political-legal 1. 2. 3.	Socio-cultural 1. 2. 3.
	Important trends, probability of occurrence			
Clients				
Employees				
Creditors				
Prof/Trade Orgs.	*Impacts on the organization, timing, threats and opportunities*			
Suppliers				
Stockholders				
Public Interest Groups				
Others				

Figure 7-1: Trends screening matrix. Source: Wallace Futures Group, LLC.

2. *Identify key stakeholders.* The next step is to identify the key stakeholders, those that have significant influence on your business. To accomplish this it is necessary to understand the business environment of the markets served by your firm. For this task, models such as Michael Porter's Five Forces Model can assist in identifying the significant market forces as well as the interplay of the key stakeholders.[8] See figure 7-2.

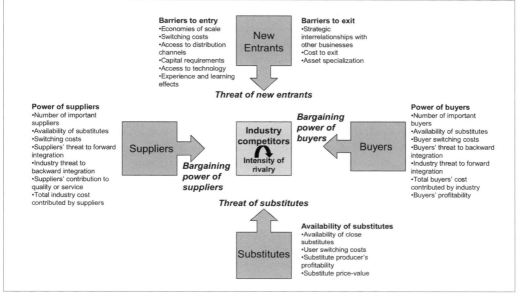

Figure 7-2: Determining important stakeholders using Porter's "Five Forces" model. Adapted from: Michael E. Porter, *Competitive Strategy,* The Free Press, New York, 1980.

3. *Locate sources of information on trends.* Drawing from observations, journals, newspapers, contacts, and so on, identify the trends that could affect your key stakeholders.

4. *Characterize the trends.* Further define the trends in terms of probability of occurrence, impacts on key stakeholders, and potential threats and opportunities. See figure 7-3.

5. *Assess the trends in terms of impact and immediacy.* Determine which trends should be addressed in order of priority. Determine threats and opportunities. See figures 7-3 and 7-4.

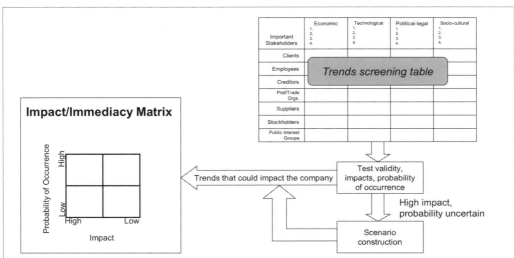

Figure 7-3: Screening trends, creating an impact/immediacy matrix. Source: Wallace Futures Group, LLC.

Converting issues into strategies and actions

Early identification of trends and issues offers little advantage unless the firm takes timely and appropriate action. Often it is very difficult for a firm to act, basically for two reasons:

1. The firm does not possess an environment for innovation, which would enable it to draw upon its full array of resources to identify, screen, and track trends and issues.

2. The firm does not have a way to process trends and issues effectively, that is, translate them into threats and opportunities and convert them into actions that create competitive advantage.

Innovation and its discontent

What makes new trends and issues hard to discern, particularly in the early stages, is that they generally do not follow established marketplace patterns. More than likely they will not appear nor take shape in a firm's established lines of business, tending instead to gather outside the normal areas of inquiry. To be able to see these trends early enough to take timely and appropriate action, it is essential to create an organization that can freely wander through markets and disciplines, seeking out patterns, early trends, and other information that, when aggregated together, may turn into important issues for the firm.

Unfortunately for many firms, this essential freedom of inquiry is stifled by its own culture and by its reward and recognition system. The pathology is relatively simple. The leaders of the firm's individual business units receive monetary rewards and advancements based primarily (or as a practical matter, solely) on the revenue and profitability earned in its defined markets. Unless they emerge entirely within the confines of the business unit, new issues tend to be ignored, since spending any time on them may only create opportunities for others and disrupt the status quo. Needless to say, this behavior is counterproductive, as it sets up destructive competition among business units at the expense of the overall success of the firm.

This ailment appears more often in the larger firms. In the small and midsized firms where resources are scarce, trust and collaboration become necessities for survival. Projects are delivered using the available people regardless of which office, business, or discipline the person happens to work in.

To enable the firm to grow and diversify, taking advantage of new markets and opportunities, the leaders of the firm must create an environment for innovation, one that fosters entrepreneurial inquiry across disciplines, business units, and geographies. Creating such an environment for innovation requires the following conditions:

- *Open organizational borders.* The organization has the ability to share or acquire information freely across organizational boundaries and the ability to draw upon the skills and expertise of the entire firm when needed.

- *An atmosphere of trust.* The organization has confidence in and reliance on the good qualities of colleagues in the firm. It also holds the belief that all are placing the good of the firm ahead of individual or business unit success.

- *Diversity.* The organization encourages diversity of backgrounds, ideas, and opinions as a way of fostering innovation.

- *Reward and recognition of innovative behavior.* The organization has a reward and recognition system that truly values openness, trust, diversity, and innovation.

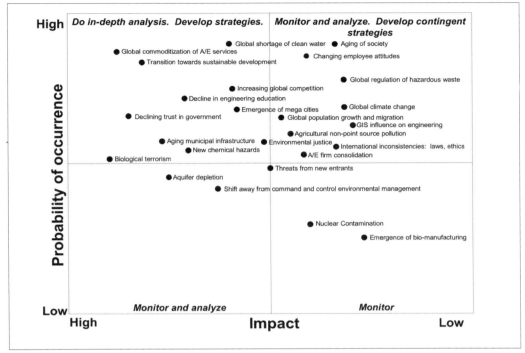

Figure 7-4: Impact/Immediacy Matrix: Examples of significant trends and emerging issues in the environmental engineering arena, along with corresponding strategies for addressing opportunities and threats. Source: Wallace Futures Group, LLC.

Finding the game changers

Most firms have some form of strategic planning process, either formal or informal. It is a process in which the firm's leadership sets the organization's future scope and direction based on past performance, current market and business conditions, and anticipated trends and issues important to the firm's future performance. Noting that trends and emerging issues form the basis for future opportunities and threats important to the firm's success, firms generally have a way of processing these trends and issues and taking action based on priority.

While this approach is useful, it does not go far enough in identifying and acting upon those trends and issues that can generate strong growth, diversification, and differentiation from the competition. In fact, simply prioritizing trends and issues and attaching some terse follow-up action may actually work to the organization's detriment, deadening the organization's senses to truly great opportunities or, conversely, firm-killing threats.

As I thought about my experience in tracking trends and issues over many years, it struck me that every so often—perhaps once or twice in a decade—issues will emerge that have such high impact that they are "game changers." They can change an industry's basis of competition. Firms that can discover these trends and act quickly and decisively can create a powerful and sustainable competitive advantage and capture significant market share.

Hazardous waste cleanup: a recent game changer

A recent example of a game-changing issue in the environmental engineering arena was the matter of the cleanup of abandoned and uncontrolled hazardous waste sites. In the mid-1970s, government environmental agencies around the United States were becoming more engaged in pollution control. Congress had passed the laws governing air and water pollution and was contemplating the regulation of municipal and industrial waste. Unlike air

151

and water pollution, these kinds of solid wastes were generally buried in the ground. There, the regulators feared, these wastes, left untended, could release contaminants and eventually pollute ground- and surface waters. Industrial wastes posed the greatest threat, since they normally contained substances that were toxic or otherwise hazardous to public health and the environment.

Environmental agencies had known about these potential problems for some time and had sought to close what they perceived as a regulatory gap. In 1976, the U.S. Congress passed the Resource Conservation and Recovery Act (RCRA), a law giving the federal government and the states authority to control the generation and disposition of so-called hazardous wastes. At the time, RCRA was thought to be preventive, that is, dealing with an issue that, if left unregulated, could cause future environmental problems.

However, as the U.S. Environmental Protection Agency began to write the regulations needed to identify and control hazardous wastes, it made an important and unexpected discovery. The problems that it sought to prevent were already there. Past industrial waste disposal practices had created thousands of sites around the country, each containing substantial quantities of hazardous industrial wastes. Many, if not most, were leaking their contents into nearby streams and aquifers, causing widespread contamination and jeopardizing public health. Unlike air and water pollution from point sources, this contamination could expose the public unknowingly to toxic chemicals. The huge public outcry that followed the discovery of hazardous waste sites like Love Canal, the Valley of the Drums, Times Beach, and other chemical disposal sites led to the passage of a law commonly referred to as Superfund.

The Superfund law provided funds and authority to investigate and clean up contaminated sites. While many of the engineering firms recognized the opportunity, few were able to muster the broad expertise necessary to perform the work. The successful firms recognized early that underground investigation and cleanup of soils and groundwater involved a number of different disciplines in science and engineering. They immediately sought to acquire the necessary resources and expertise to respond to the opportunity. Those firms that failed either did not recognize the opportunity or tried to address it within their existing disciplines. Many called the hazardous waste contamination issue a "flash in the pan," failing to recognize the large toxics issue that loomed large in the mind of the public. Table 7-1 lists other examples of game-changer issues past, present, and future.

Table 7-1: Examples of game changers: past, present, and future. Source: Wallace Futures Group, LLC.

Issue, Trend, Event	Description	Impact	Timing
Past Game Changers			
Decline of the nuclear power industry	Concerns over nuclear power stopped plant development	Loss of nuclear plant siting, engineering business	Medium
Passage of environmental laws National Environmental Policy Act, Clean Water Act, Clean Air Act	New laws applied environmental regulations to business	New opportunities in environmental business	Medium
Passage of Superfund	Unprecedented authority given to force and fund hazardous waste cleanups	New opportunities in hazardous waste remediation	Fast
"Environmental Olympics"	Environmentalism included in selection criteria for cities hosting the Olympics	Entry point for engineering firms to support candidate cities	Medium

Table 7-1: Examples of game changers: past, present, and future. (continued)

Issue, Trend, Event	Description	Impact	Timing
Game changers, present and future			
Sustainable development.	New concerns form over consequences of development	New opportunities in broad spectrum of environmental projects	Medium
Events of September 11	Unforeseen threats and vulnerabilities to open systems and infrastructure appear	Sea change in the design of systems and infrastructure	Fast
Globalization	Trade barriers drop everywhere; new forms of commerce appear	New opportunities, new competitors everywhere in the world, 24/7	Fast
Corporate transparency and accountability	Stakeholders demand openness following events of Enron, WorldCom	Requirements for comprehensive performance reporting	Medium
Knowledge management and virtual collaboration	Information technology enables new forms of knowledge sharing and access	Access to knowledge and to people to work together electronically	Medium

What about sustainable development?

Is sustainable development a game-changing issue? In my opinion, the answer is most definitely yes. Although the movement will not emerge as fast as Superfund did, sustainable development has all of the ingredients of a game-changing issue. As stated in chapter 2, the movement toward sustainability will force a substantial overhaul of the current production-consumption infrastructure. Such an overhaul will require the invention, development, and deployment of new technologies, processes, and systems, many of which have yet to be invented. Furthermore, their development and deployment will be accomplished under increasing public scrutiny, ensuring that the overhaul will not compromise sustainability or quality of life objectives. Applying new technologies, processes, and systems to create a sustainable infrastructure is nominally the work of engineering firms. However, success in developing and deploying these items will depend largely on the ability of the engineering firms to assemble the requisite knowledge and skills in sustainable development and to engage effectively those key stakeholders affected by the change. Firms that can develop those capabilities and can engage credibly with key stakeholders will have a distinct competitive advantage over firms that cannot.

Developing a "futures watch" process

For many organizations, making projections of what trends and events will be important in the coming years seems more like fortune-telling than a meaningful business activity. Yet for strategic planning, developing a set of possible future conditions is just as essential a task as competitor analysis or client needs assessments. Therefore, it is important for companies to have a sound program for thinking about the future, that is, a systematic way of determining the important trends, issues, and events that will likely affect both the organizations and their stakeholders within the organization's planning horizon. Unfortunately, futures analysis and forecasting, the process for distilling a set of salient futures out of myriad possibilities, is not normally found in most organizations' skill sets.

Recognizing this shortfall, many organizations turn to consulting futurists, domain experts, or other analysts to help them understand what the future may hold for their companies. In addition, the organizations may use these individuals to help them explore matters outside the confines of their everyday businesses. By "stepping outside the box," organizations hope to get a handle on new possibilities that may be obscured by their participation in day-to-day operations. Through this sort of exploration, organizations may find and take advantage of new opportunities before their competitors do. They may also identify and take steps to mitigate serious threats before they cause problems.

Still, identifying opportunities and threats is only a first step. Unless the organization has a method for processing potential opportunities and threats and converting them into actions, the results of exercises in thinking about the future will be wasted. To these ends, this section proposes a systematic approach for capturing and processing trends, events, and issues. This approach has three components:

- *Establishing your comfort zone.* What are your established procedures and trusted sources for gathering and processing information, and taking action on potential opportunities and threats?

- *Setting up a "futures watch" process.* How could you revise your current process to be more effective in making decisions while maintaining overall an acceptable level of risk?

- *Identifying and taking action on new opportunities and threats.* How would you operate a "futures watch" process and connect it to your strategic plan?

Establishing your comfort zone

Engineering firms attract many bright people with good observational skills and market insights. Furthermore, these employees are not bashful in offering ideas about potential new opportunities and threats to their managers. However, it is rare that these ideas are ever properly assessed or developed in a timely manner. The reason that most firms fail to capitalize on these ideas is twofold: (1) they do not have an established process for collecting and processing these ideas and (2) they have neither clearly defined nor communicated the level of risk the firm's management and leadership are comfortable in taking.

Most of the "let's think about the future" strategy sessions I have attended, while exhilarating, have not been especially productive. This may sound blasphemous coming from an author who has spent many years arguing for more "futures thinking" in firms. However, in these sessions, the facilitators usually begin with the premise that session success is predicated on the ability of the participants to stretch their thinking well outside their normal bounds. In so doing, the facilitators argue, the participants can identify new market opportunities, find new business processes, strike up new business relationships, and create new operating paradigms. Although these ideas may offer tremendous benefits to the firm, they need to be screened and processed in a way that takes into account the firm's appetite for risk and change. In other words, no matter how intriguing or beneficial the idea may be, it will go nowhere if it doesn't fit within the firm's comfort zone. This is the zone that takes into account the institutional stewardship and fiduciary responsibilities of the firm's leadership, that is, making sure that ventures into new market and operational territories are balanced against long-term profitability and success.

Fortunately for the firms that have been in business for some time, their leaders have a good sense of their comfort zone. Unfortunately, most have not communicated those parameters to the rest of the organization, nor have they set up procedures for cultivating new ideas within the zone. Consequently, many opportunities are missed, having never been shown to be compatible with the firm's sense of acceptable risk or its ability to initiate and manage change.

How do you find your comfort zone? Here are a number of questions that help to locate an organization's comfort zone:

- *What is your business, honestly?* Think about how you would describe your business honestly to a trusted friend. Who are your clients and what issues do they face? What is your business environment? What are your top issues and concerns, that is, what are the matters that distract you the most? What are the significant opportunities and threats you expect to face?

- *Where do you get your information?* What are your sources of information about your business: its markets, financial situation, resources, trends, issues, and other important matters? How is the information collected, processed, and communicated? Is any systematic analysis done on this information? How do you validate the information you rely on to manage your business? How well do you use your clients as either sources of information or validation of assumptions? Overall, how well do you trust your sources of information and your means of validation?

- *What are your procedures for processing information?* What systems, processes, and tools do you use to analyze the information gathered? Do you have a trusted group of people, both inside and outside the firm, to whom you can go to discuss trends and issues, test assumptions, and explore new ideas? In turn, do these sources have additional networks that you can tap for additional information or analysis? Here your clients are your most important test bed for new ideas, since they determine how trends and issues may turn into actual projects. However, care must be taken in factoring client opinions into your plans. Clients generally operate in a single domain and, consequently, have a relatively narrow perspective. Engineering firms operating in multiple sectors should take advantage of their broad perspectives, regularly engaging their clients in discussions about what is happening in other industry sectors.

- *What is your response threshold for opportunities and threats?* What does is take in the form of convincing information to get your organization to invest in a new idea or opportunity? What are your levels of response, for example, monitor and report, pilot studies, new ventures? Do you have an annual budget set up to fund these responses?

Establishing the process

Based on the answers to the questions posed above, an organization can begin to formulate a set of procedures for identifying and assessing trends, issues, and events as they might pertain to its current and future success.

- *What is your past history of success?* Think of some examples in which your organization identified a new opportunity or a threat. How did the organization respond to the opportunity or threat? Did the organization act sufficiently and in a timely manner? Who championed the idea? How much of a struggle was there for this champion to raise the matter to management? Would he or she do it again?

- *Should you revise your current process?* Should you bring more or less attention to the identification and processing of new ideas? Do you need new sources of information? Are your current sources sufficiently knowledgeable and trustworthy? Should you bring new participants into the decision-making process? Do you need new processes and tools? Does your organization have an environment for innovation, as defined earlier in this chapter? Do you need to update your investment decision criteria? At what frequency should you review progress?

Exploring the future

Once you are comfortable with your process revisions, you should energize the system and begin to explore and analyze trends, issues, and events within your planning horizon. It is important to remember that information about impending changes, opportunities, and threats initially present themselves as weak signals.

- *"Pulse" your sources.* Even though you may have amassed a sufficient quantity of trusted sources of information relative to your business, you should not rely on them to bring the information to you. Instead, you should set up a routine for contacting them at some predetermined frequency, or immediately when an issue arises.

- *Conduct regular assessments.* It is easy for an organization's futures exploration process to become subordinated by pressing schedules and events. Therefore, it is important for the organization's leadership to review and assess progress made on its current portfolio of initiatives.

- *Fine-tune your decision-making process.* In addition to the progress assessment, the leadership of the organization should review the process and results of past decisions to determine the effectiveness of the decision-making process. One trap to avoid is the resistance of the existing organization to new initiatives and change, often prompted by threats to budgets or the existing organizational structure. Another trap is the reluctance of the champion to stop pursuing his or her idea, even though it is unlikely to prove successful. Here the organization must set up clear "go/no go" policies and criteria for several stages of exploration.

> **Key trends, issues and events driving the adoption of sustainable development**
>
> - The occurrence of a major environmental or resource disaster: oil spills, nuclear disasters, ecological losses, ecoterrorism
> - Detection and assessment of risks associated with toxins in the environment
> - Development of remote sensing and assessment technology
> - Availability of freshwater and sanitation
> - Quantitative and qualitative determination of the status of critical resources: renewable and nonrenewable
> - Development of alternative energy sources: solar, wind, hydrogen
> - Emergence of green certification programs
> - The way the United States government chooses to manage the climate change issue
> - The progress and acceptance of genetic engineering
> - Results of the Millennium Ecosystem Assessment
> - The development of megacities

What is the future of the sustainable development movement?

For the leaders of engineering firms who are thinking about building a business related to sustainable development, a number of obvious but legitimate questions come to mind:

- What does the future hold for sustainable development?

- What do your clients think about sustainable development today? How will they think about this issue three, five, or 10 years from now?

- What services will your clients need relative to sustainable development?

- How can you and your firm prepare to deliver these services?

- What events might delay, speed up, or change the movement toward sustainability?

While I certainly have no crystal ball that can foretell the future, I can lay out what I believe to be a likely scenario of the future of sustainable development as it relates to the engineering practice. In this scenario, I have attempted to look forward by five, 10, and 20 years, characterizing each step forward with what I think will be the prevailing theme for business. For each time period, I have noted important trends and market drivers that will create opportunities and threats. Also noted are key events that could alter these futures. Last, I offer some suggestions on how firms might position themselves to take advantage of the opportunities or, conversely, avoid the threats, that will emerge in these time periods.

Given the uncertainty in today's world, there are clearly a number of plausible and perhaps reasonable futures equal to the one I am proposing. Indeed, attempts by companies to plan their future based on a single scenario usually end in failure. However, my approach is not to argue for a single future upon which to base all planning. My approach is to offer up a likely future based on current trends that companies can use as a point of embarkation. From there, they can steer a course using the scenario as a guide, while paying close attention to key trends and events that will shape the market and influence key stakeholders.

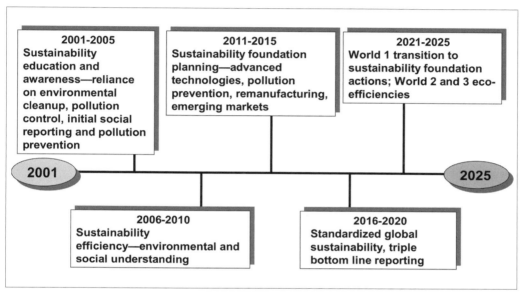

Figure 7-5: Sustainable development: future timeline: Source: Unpublished report by Coates and Jarratt, "Sustainability in the 21st Century, Implications for Business 2001-2025, Business Case for Sustainability Briefing," Spring 2002.

2004–2009: Educating industry and government

In the United States, sustainable development education and awareness building will be the key themes over the next five years. Today sustainable practices in companies are primarily reputation driven, caused by pressures from nongovernmental organizations (NGOs) and other stakeholders. Most major corporations, particularly the multinationals, have recognized that their ability to operate as a global corporation is based on stakeholder perceptions of their environmental and social as well as economic performance.

At this juncture, image and reputation stand out as the primary reason for organizations, both private and public, to make public commitments to sustainability. Responding to enhanced demands for transparency and accountability, they have built a business case for sustainable development around this theme. However, inside and outside this group, a number of organizations are learning that following the precepts of sustainable development can be a source of innovation, providing new and novel ways of reducing costs and increasing both revenue and profit.

2004–2009: Education and awareness building on the benefits of sustainable development practices
In this time period, the emphasis on sustainable development is primarily stakeholder driven, pushing companies to make their operations environmentally friendly, energy conserving. There will be increased emphasis on brownfield cleanup, redevelopment of urban areas, pollution prevention, energy efficiency. Corporations will be required to become more transparent and accountable.

Trends and market drivers
• Increasing pressures from NGOs, citizen groups on corporate image and reputation. They will push for sustainable development reporting.
• Corporations, government agencies will seek to improve their environmental and social practices. They will demonstrate organizational commitment through green buildings, ecolabeling, and pollution prevention.
• Interest in green products and services by market segments, e.g., "cultural creatives,"[9] will increase.
• Turmoil in the Middle East will distract policy makers from sustainable development concerns. However, as oil supplies become vulnerable, there will be an increasing interest in renewable energy.
• Concerns about homeland security will instigate a movement toward decentralizing utilities, e.g., energy, water, sanitation, to make the infrastructure more robust.
• Economic conditions – funds available for infrastructure development, rehabilitation – are an important consideration.

Preparation
• Develop a comprehensive understanding of sustainable development issues.
• Develop or acquire capabilities in sustainable development reporting, LEED certification, energy efficiency, pollution prevention, life cycle analysis.
• Reanalyze core markets, especially those composed of client groups with large environmental footprints, sustainability interests.
• Discuss sustainable issues with clients and assess their position and interest. Focus on innovation and economic benefits.
• Develop relationships with key stakeholder groups in clients' domain.

Trends to watch
• Economic conditions.
• Progress on resolving the Middle East conflicts: Iraq, Afghanistan, Israel, Palestine.
• Legislation on urban sprawl, global warming; development of hydrogen technology.
• Client interest and commitment to sustainable development.
• Growth of ecolabeling.
• Growth of sustainable cities initiatives.
• National, state initiatives on sustainable development.
• Emergence of new, more-sustainable technologies.
• Results of the Millennium Ecosystem Assessment.
• Emphasis, progress on fulfillment of commitments to Millennium Development Goals.
• Success of multinational corporations that have made commitments to sustainability, e.g., Dow, DuPont, Ford, STMicroelectronics.
• Market success of sustainable products, e.g., hybrid autos, LEED-certified buildings.

Wild cards
• An environmental Enron or WorldCom.
• Significant events of ecological or resource degradation.
• Another significant terrorist attack on U.S. soil.

2010–2015: Transition toward sustainability

At this point, there is strong public opinion that its quality of life is unsustainable. Furthermore, unless something is done to change, things will get worse. Consumers are taking these concerns to the marketplace and buying products and services from companies that can demonstrate their commitment to sustainability. Some of these opinions are being heard by legislators and are being translated into laws and regulations.

2010–2015: Transitioning toward sustainable development practices
Drivers for sustainable development shift from stakeholders to market forces. The marketplace now recognizes the economic benefits of sustainable development. Use of life cycle analysis is commonplace. Corporations engage in research and development (alone or through industry and government partnerships) to develop new energy technologies, substitutes for critical, nonrenewable resources. Eco-efficiency and eco-effectiveness have become embedded in industry design practices. Product durability, repairability replaces product obsolescence. Environmental legislation, product and material specifications, design standards are amended to incorporate additional recycling and reuse. In 2012, the "Johannesburg plus 10" summit reports significant progress made toward the Millennium Development Goals.

Trends and market drivers

- Increasing pressures from stockholders, customers, other stakeholders for incorporating sustainable development principles in design and operations to achieve better financial performance.
- Emergence of new forms of project procurement based on life cycle performance.
- Increasing corporate transparency through advances in information technology, telecommunications improves information access and exchange among NGOs, customers, shareholders.
- Results of the Millennium Ecosystem Assessment set additional priorities for environmental and social improvements.
- New and more sustainable technologies and processes emerge. Technological advances in satellite imaging, sensors, etc., improve ecological and resource assessment capabilities.
- Demand for green products continues to rise as customers realize the benefits of buying more-durable, recyclable products made from less-toxic, renewable materials.
- New markets for sustainable products and services emerge in the developing world.
- Economic conditions continue to influence the amount of funds available for infrastructure development, rehabilitation.

Preparation

- Develop an acute awareness of how sustainable development issues affect clients in core markets.
- Develop top-notch capabilities in the application of sustainable development technologies and systems on client projects.
- Be capable of full-service project delivery for projects that include aspects of sustainability in the scope of work.
- Actively track the performance of other projects on sustainable development dimensions. Know what others have achieved.
- Research opportunities in the developing world. Position your company to be able to deliver services overseas in the developing world.
- Assess prior performance on sustainability projects and compare with competitors.

Trends to watch

- Economic conditions.
- Policies, actions taken by clients related to sustainable development.
- Policies, legislation, actions by federal agencies, states, local governments.
- Preparations for the "Johannesburg plus 10" summit in 2012.
- Progress accomplishments and disappointments and their relation to the Millennium Development Goals.
- Continuation of the activities in the Millennium Ecosystem Assessment.
- Scope and extent of corporate sustainable development reporting. Importance of triple bottom line reporting to financial markets.
- Development of sustainable technologies, environmental sensors, risk assessment tools.
- Scope and level of investment in developing countries by governments, international funding agencies.
- Continued advances in sustainable technologies, processes.

Wild cards

- Terrorist attacks.
- Significant events of ecological or resource degradation.
- Technological breakthrough in energy production.

2016–2025: Shifting the focus to the developing world

In this time frame, the United States as well as other World 1 nations have become more in tune with the problems of the developing world. They recognize the quality of life aspirations of Worlds 2 and 3. In addition, they recognize that unless some intervention is made, their development will be achieved through the use of nonsustainable technologies. International funding agencies help to facilitate the transition to and the use of sustainable technologies. Corporations continue to mine previously untapped markets by developing new products and services that meet the special needs of the developing world.

2016–2025: Shifting the focus of sustainable development from World 1 to World 2 and World 3
Market forces for sustainability in the developed world have taken hold. Emphasis shifts to the developing world, which now holds most of the resources and production capacity. For the most part, the developing world agrees that sustainable practices ought to be employed, but it prefers growth and economic development over sustainable development, that is, waiting for sustainable technologies to become available. Multinational corporations continue to apply World 1 standards to their operations across the world. In Worlds 2 and 3, international funding agencies are investing in the cleanup of past industrial practices and in helping these countries protect their environment and resources. Developments in nanotechnology, (MEMS) Micro-Electro-Mechanical Systems, sensors, and imaging technology now make it possible to detect and trace the movements of materials and products across the world.

Trends and market drivers

- Continued corroboration that past warnings of nonsustainable development were well founded and efforts to mitigate are paying off.
- Evidence of ecological overshooting in some areas such as commercial capture fishing, dead zones in some sections of oceans, certain megacities, and destitute countries.
- Development of efficient cleanup technologies.
- Continued advances in nanotechnology, MEMS, sensors, and satellite imaging technology.
- Improvements in governance, reduced corruption in World 2 and 3 nations.
- Continued development of the technologies necessary to support eco-efficiency, eco-effectiveness.

Preparation

- Develop capabilities to deliver projects and conduct business in selected developing countries.
- Develop relationships with international funding agencies, international associations, multinational companies.
- Track new technologies: nanotechnology, MEMS, sensors, and imaging, remediation.

Trends to watch

- Economic conditions.
- Programs to identify and remediate contamination in World 2 and 3 countries.
- Improvements in the framework conditions necessary to conduct business in developing countries.
- Preparations for the "Johannesburg plus 20" summit in 2022.
- Success of companies in marketing to the "other 5 billion" people.
- Advances in sustainable technologies. Successes or failures in application.

Wild cards

- Disaster in biotechnology product.
- Another Bhopal or Chernobyl-like event.

Conclusion

The assessments captured in these three time frames project a continuing advance sustainable development as a most significant future issue. They support the thesis of this book: sustainable development is a new and important issue that is a sensible and ethical imperative as well as a market opportunity for the engineering community. Engineering firms may ignore this trend at their peril.

The question before us is how fast to react. Engineering firms earn their income by managing projects and relationships with a wide variety of clients from both the public and private sectors. These clients reside in many locations, are subject to many laws and regulations, operate in a variety of cultures, and deal with many different stakeholders. Thus, the engineering community's reaction to sustainable development needs to be tempered by the nature of its clients. There is no "one size fits all" answer, but rather a set of responses based on each client's circumstances.

It seems, therefore, that the first order of business ought to be education: helping clients and their stakeholders understand their place in a larger picture of the world. Again, engineers are uniquely qualified to provide this educational component. How soon they can come together and develop the educational message is up to them.

---

[1] Anonymous Chinese proverb, from *Columbia World of Quotations* (New York, NY: Columbia University Press, 1996).

[2] Corporate Social Responsibility Newswire Service (CSRwire), http://www.csrwire.com/

[3] *Industry as a Partner in Sustainable Development: Consulting Engineering* (Geneva, Switzerland: International Federation of Consulting Engineers, 2002).

[4] Robert Cassidy, "What Next for the Sustainability Movement?" *Building Design and Construction*, http://www.bdcmag.com/contactbdc/sustainableedit.asp.

[5] Michael Harris of CH2M HILL, in discussion with the author, 1999.

[6] Cassidy, "What Next for the Sustainability Movement?"

[7] World Future Society, *The Future: An Owner's Manual* (Bethesda, MD: World Future Society, 2002), 2–3.

[8] Michael Porter, *Competitive Strategy* (New York: The Free Press, 1980).

[9] "Cultural creatives" is a term coined in a book of the same name by Paul Ray and Sherry Booth Anderson, to describe the emergence of a new culture of people who care deeply about ecology and saving the planet, about relationships, peace, social justice, and about self actualization, spirituality and self-expression. Surprisingly, they are both inner-directed and socially concerned, they're activists, volunteers and contributors to good causes.

APPENDIX A

Useful Sustainable Development Publications

General Philosophy and Strategy

***Our Common Future*, The World Commission on Environment and Development, Oxford University Press, Oxford, UK, 1987.**

The report by the commission chaired by Gro Harlem Brundtland on the state of the world and the connections between the economy and environment. This is the report that coined the phrase sustainable development.

Evaluation: The basis for the world movement toward sustainability.

***The Ecology of Commerce: A Declaration of Sustainability*, Paul Hawken, Harper Business, New York, 1993.**

The farsighted work of Paul Hawken, one of the founders and true visionaries on the sustainable development movement.

Evaluation: A must-have for any sustainable development library.

***Technology, Humans, and Society: Toward a Sustainable World*, Richard C. Dorf, Academic Press, San Diego, CA, 2001.**

A sweeping discussion from many contributors of almost every aspect of sustainable development, mostly from the scientific, engineering, and business perspectives.

Evaluation: A must-have resource for sustainable development.

***Eco-Economy: Building an Economy for the Earth*, Lester R. Brown. W.W. Norton & Company, New York, NY, 2001.**

A cry for harmony between the economy and nature. Author Lester Brown uses examples of past failed civilizations to show that our current take-make-waste economic model is destroying the support systems on which our model is based. To make his point, Brown extrapolates that if China were to follow the U.S. model for automobile use (e.g., at least one car per family), China would need 80 million barrels of oil a day, more than the world's current oil output. He closes with a discussion on how to make the transition to sustainability.

Evaluation: Vintage Lester Brown. Well written with plenty of supporting data. His path forward is idealistic, but certainly worth the time reading and thinking through his premises and approach.

***Earth at a Crossroads: Paths to a Sustainable Future*, Hartmut Bossel, Cambridge University Press, New York, 1998.**

The book begins by noting the shift from a local-regional society to a global society and all its attendant problems. The author looks at the world as a complex and dynamic system, which, if poorly tended, could result in incredible disasters for future generations, the basic premise for sustainable development. He debunks the economist's logic of discounting the future, noting that this logic discounts the interests of future generations. Sustainable development does not allow discounting the value of a future ecosystem. Part 2 of the book examines two alternative development paths: Path A (competition) and Path B (partnership). In part 3, the author determines that Path A is not sustainable.

Evaluation: More philosophical than practical, but a good resource for discussing the intellectual basis for sustainability.

***Making Development Sustainable: The World Bank Group and the Environment—Fiscal 1994*, The International Bank for Reconstruction and Development / The World Bank, Washington, DC, 1994.**

An older book that describes the shift in World Bank policy in its design of development policies and investment. In 1987, the World Bank started an effort to incorporate environmental concerns into all aspects of its work.

Evaluation: Good for reviewing World Bank projects in the early 1990s and seeing how environmental factors were considered.

***Envisioning a Sustainable Society—Learning Our Way Out*, Lester W. Milbraith, State University of New York Press, Albany, 1989.**

An analysis of the state of the world: how we have come to this point, why science and technology will fail to solve these problems, and how we as a society must change to avoid ecological catastrophe. Although over 10 years old, the publication is still referenced by other authors.

Evaluation: Highly philosophical but useful in its discussions of the evidence of nonsustainability. References to the Club of Rome study and the tragedy of the commons.

***Harvard Business Review on Business and the Environment*, Harvard Business School Press, Boston, MA, 2000.**

A collection of Harvard Business Review articles regarding the relationship of business to the environment and sustainability. Some are quite old, for example, from 1991.

Evaluation: Supports the case that sustainability has become accepted in one of the leading business schools.

***You Can't Eat GNP: Economics as Though Ecology Mattered*, Eric A. Davidson, Perseus Publishing, Philadelphia, PA, 2000.**

Enriched by the author's personal stories, this book delivers a "bottom-up" message supporting the concept of natural capitalism. He also challenges the message of the technology-optimists: "don't worry, be happy, technology will save the day."

Evaluation: Easy reading but with powerful arguments about the importance of natural capital and the fallacy of using gross national product as the primary indicator of successful growth and development.

***Earth Rising—American Environmentalism in the 21st Century*, Philip Shabecoff, Island Press, Washington, DC, 2000.**

A well-written discussion on the state of environmentalism today by the science writer of the New York Times.

Evaluation: Fascinating reading. Should be read by anyone engaged in sustainable development issues.

***Financing Change*, Stephan Schmidheiny and Federico J. L. Zorraquin with the World Business Council for Sustainable Development, The MIT Press, Cambridge, MA, 1996.**

This book seeks to answer the question, "Are the world's financial markets … a force for sustainable human progress or are they an impediment?" The authors attempt to answer this question by investigating how the various participants' company leaders, investors, bankers, insurers, accountants, and raters—respond to the environmental and social demands of sustainability.

Evaluation: A sound treatise that examines the key issues related to how the financial markets affect progress toward sustainability.

In Earth's Company: Business, Environment and the Challenge of Sustainability, Carl Frankel, New Society Publishers, Gabriola Island, BC, Canada, 1998.

In this insightful book, the author describes the first three eras of corporate environmentalism—compliance, disclosure, and beyond compliance. In the author's view, all of these were well meaning (well, sort of) but too narrowly focused. He then describes the fourth era, whole systems thinking.

Evaluation: An essential read for people intending to work in the sustainable development field.

The Sustainable Business Challenge—A Briefing for Tomorrow's Business Leaders, Jan-Olaf Willums with the World Business Council for Sustainable Development, Greenleaf Publishing, Sheffield, UK, 1998.

Written in the form of a string of memorandums, letters, news items, and journal articles—an imagined dialogue within a hypothetical company—the book walks the reader through the salient principles and issues surrounding sustainable development.

Evaluation: Engaging work discussing the business case for sustainable development.

Mid-Course Correction, Ray C. Anderson, The Peregrinzilla Press, Atlanta, 1998.

Ray Anderson is founder, chair, and CEO of Interface, Inc., the world's largest producer of commercial floor coverings. This book recounts Anderson's career and his journey to his sustainable vision for his company. In the book, he poses the Interface model of corporate sustainability, describing it as the prototypical company of the 21st century. He also provides company statistics showing that it is possible to do well by doing good.

Evaluation: If you haven't read or heard Anderson speak about his career and how he learned about sustainable development, this is a good book to get. His model of the prototypical company of the 21st century is also worthwhile.

Natural Capitalism, Paul Hawken, Amory Lovins, and L. Hunter Lovins, Little, Brown and Company, Boston, 1999.

The authors make operational John Elkington's notion of a triple bottom line. When economists discuss capital, they usually mean financial capital: monies invested to produce commercial goods and services, and a return on the investment. But there is another branch of economists—ecological economists—who recognize three other types of capital: human capital, manufactured capital, and natural capital. Human capital consists of the aggregate knowledge, skills, and competencies of individuals within an organization or institution. Manufactured capital refers to the physical aspects of capital, that is, the factories, buildings, dams, roads, ports, and other humanmade improvements. Natural capital refers to the non-renewable resources (minerals, metals, and fuels), and the renewable (ecosystems). Nonrenewable resources supply many of the essential materials. Industry uses the first three types of capital to transform the fourth into the products and services of our daily lives.

Evaluation: A very readable book by perhaps the three most knowledgeable people on sustainable development. Many examples of how engineers and designers have been extremely narrow minded in their approach to the design of facilities and systems.

Developing a Business Case

Walking the Talk: The Business Case for Sustainable Development, **Chad Holliday, Stephan Schmidheiny, and Philip Watts, Berrett-Koehler Publishers, San Francisco, 2002.**

The authors, three leaders of large multinational companies, tell how and why they became convinced that the sustainable development movement is important to their companies. With the help of the World Business Council for Sustainable Development and its members, the authors show what they call the 10 building blocks of sustainability. Each building block chapter contains supporting case studies contributed by member firms.

Evaluation: A well-written book containing valuable source material for developing the business case for sustainable development.

Changing Course: A Global Business Perspective on Development and the Environment, **Stephan Schmidheiny, MIT Press, Cambridge, MA, 1992.**

Business's response to the Brundtland Commission report, termed as a defining moment in the history of corporate environmentalism. Extensive discussion on the problems and issues facing business in the time following the commission's report. Contains many case studies of companies taking steps to implement sustainable development principles in their organizations.

Evaluation: An important milestone in the business movement toward sustainability.

The Natural Step for Business: Wealth, Ecology and the Evolutionary Corporation, **Brian Nattrass and Mary Altomare, New Society Publishers, Gabriola Island, BC, Canada, 1999.**

Authors Brian Nattrass and Mary Altomare describe a new form of business model based on the Natural Step (TNS) principles. The book starts with chapters on setting a new framework for business and follows with a series of case studies on four companies. Part 3 provides lessons, tools, and methodologies.

Evaluation: A clear, concise description of the application of TNS principles to business.

Dancing with the Tiger: Learning Sustainability Step by Natural Step, **Brian Nattrass, and Mary Altomare, New Society Publishers, Gabriola Island, BC, Canada, 2002.**

Follow-up to the authors' previous book, The Natural Step for Business. In this latest book, authors Nattrass and Altomare focus on four companies, Nike, Starbucks, Whistler Resort, and CH2M HILL, and relate the stories of each company's quest to achieve sustainability.

Evaluation: An interesting and well-written book replete with stories of how firms from very diverse sectors decided to incorporate sustainable development into their culture and corporate strategy.

Corporate Social Responsibility

***The Naked Corporation: How the Age of Transparency Will Revolutionize Business*, Don Tapscott and David Ticoll, Free Press, New York, NY, October 2003.**

This work documents the situation and the consequences of the prediction by the World Business Council for Sustainable Development that in the future "everyone will know everything about you all the time." The authors note the new power of public interest groups and nongovernmental organizations that is enabled by information technology. They call the force transparency, which in their minds means unprecedented access by stakeholders to all kinds of information about company activities and performance. Business Week notes that this new demand for information actually gives companies a new way to differentiate.

Evaluation: Well-developed arguments about sustainable development drivers, corporate social responsibility, and accountability written by two authors who come from a nonenvironmental area of business.

***When Good Companies Do Bad Things*, Peter Schwartz and Blair Gibb, John Wiley and Sons, Inc., New York, 1999.**

The book illustrates the new power of nongovernmental organizations in damaging corporate reputations if they observe bad corporate behavior. It provides case examples of the wake-up calls received by companies like Shell, Nike, Texaco, and Nestlé. It then addresses best practices for corporate responsibility as well as ways to incorporate these practices into long-term strategies. Peter Schwartz is well known in the areas of scenario analysis and is the author of The Art of the Long View.

Evaluation: The "wake-up call" stories are well written and illustrate the growing trends related to corporate responsibility. The book contains additional insights on converting this new responsibility into a competitive advantage.

***The Stakeholder Strategy: Profiting from Collaborative Business Relationships*, Ann Svendsen, Berrett-Koehler Publishers, San Francisco, CA, 1998.**

Recognizing the new power of public interest groups and nongovernmental organizations, the author offers a new and more collaborative approach to stakeholder relations. The book contains a six-step guide to the stakeholder relationship-building process, built along the lines of "plan-do-check-act": (1) foundation creation, (2) organizational alignment, (3) strategy development, (4) trust building, (5) evaluation, and (6) process repetition.

Evaluation: A comprehensive and strategic approach to developing stakeholder relationships.

***Corporate Social Investing: The Breakthrough Strategy for Giving and Getting Corporate Contributions*, Curt Weeden, Berrett-Koehler Publishers, San Francisco, CA, 1998.**

In this book, the author seeks to replace the haphazard efforts of corporate philanthropy with a 10-step strategy model for corporate social investment. If corporations are to be perceived as good corporate citizens and develop good relationships with their stakeholders, then they will need to move away from social investment by whim to a more strategic approach. Each social investment must be based on a sound business reason.

Evaluation: A handbook and a required read for companies that participate in corporate giving at any level.

Waltzing with the Raptors: A Practical Roadmap to Protecting Your Company's Reputation, Glen Peters, John Wiley & Sons, New York, NY, 1999.

This book is based on the realization that the relationship between corporations and their stakeholders has changed. The author sees the exponential rise in growth and power of stakeholder and activist groups and their willingness to act to right what they perceive as corporate wrongdoings. The book offers a reputation assurance framework that can help companies maintain and (hopefully) improve their reputations.

Evaluation: An engaging discussion about reputation management and working with stakeholders who, for their speed and destructive power, the author equates to velociraptors.

Sustainable Development Indicators

***Our Ecological Footprint: Reducing Human Impact on the Earth*, Mathis Wacker-nagel and William Rees, New Society Publishers, Gabriola Island, BC, Canada, 1996.**

This book presents the concept of an ecological footprint, the load imposed by a population on its surrounding ecological resources.

Evaluation: This concept is presented in a very understandable fashion supplemented with many illustrations. There is a Web site on which the reader can determine his or her personal ecological footprint: www.lead. org/leadnet/footprint/intro.htm.

***Sustainable Measures: Evaluation and Reporting of Environmental and Social Performance*, Martin Bennett, Peter James, and Leon Klinkers, eds., Green-leaf Publishing, Sheffield, UK, 1999.**

A collection of essays on sustainable performance measurement, covering environmental and social evaluation and reporting.

Evaluation: A broad and comprehensive survey of the state of the practice in environmental and social performance evaluation.

***State of the World 2003: A Worldwatch Institute Report on Progress Toward a Sustainable Society*, Gary Gardner (project director), W. W. Norton & Company, New York, 2003.**

The 20th edition of the State of the World publication. This edition contains eight chapters on existing conditions and future challenges on various subjects including biodiversity, gender gaps, energy, urbanization, mining, and religion.

Evaluation: Although seen by critics as too apocalyptic, this work is and has been a major influence on sustainable development thinking.

***Vital Signs 2003: Trends That Are Shaping Our Future*, Michael Renner (project director), W. W. Norton & Company, New York, 2003.**

Published in cooperation with the United Nations Environment Programme, this book contains information and analysis on the state of the world, presented in both chart and narrative form. The chapters in the second section of the book cover special issues with some overlap with the State of the World publication.

Evaluation: As in the case of the State of the World publications, this too is seen by critics as biased against development. The data appear to be well researched. Trends are reported in a neutral way.

***World Resources 2002–2004: Decisions for the Earth: Balance, Voice, and Power*, United Nations Development Programme, United Nations Environment Programme, World Bank, and World Resources Institute, World Resources Institute, Washington, DC, 2003. The entire publication is available at http:// pubs.wri.org/pubs_pdf.cfm?PubID=3764.**

The 10th in a series of comprehensive reports on the state of the world's ecosystems. This report defines and discusses environmental governance and its importance in reversing ecological decline.

Evaluation: Excellent resource. A first-time country by country evaluation of governmental performance in citizen access and transparency. The World Resources Institute is making its World Resources database available online in the companion Web site EarthTrends (http://earthtrends.wri.org/).

Vanishing Borders—Protecting the Planet in the Age of Globalization, **Hilary French and the Worldwatch Institute, W.W. Norton Company, New York and London, 2000.**

A look at the profound implications of accelerating globalization for our planet's health and a prescription for the action necessary to cope with this challenge. It tackles the numerous and complicated issues of globalization and the environment and makes them understandable. The book notes the increasing power of public and nongovernmental organizations derived through advances in telecommunications.

Evaluation: Sourcebook for the evidence of the effects globalization is having on the earth's resources and carrying capacity. The book also contains evidence of the positive response by industry to solve the problems.

The Heat Is On, **Ross Gelbspan, Addison Wesley, Reading, MA, 1997.**

Written as an exposé, the author details how the oil and gas industry are misleading the public about global climate change.

Evaluation: This 1997 work has been eclipsed by events. The major players in the fossil fuels industry have now relented and are working to reduce greenhouse gas emissions and moving into alternate forms of energy.

Industrial Environmental Performance Metrics—Challenges and Opportunities, **National Academy of Engineering, National Academy Press, Washington, DC, 1999.**

Noting that (1) what gets measured gets done and (2) firms increasingly view environmental performance as an area of possible competitive advantage, the National Academy of Engineering investigated the potential for using environmental performance metrics for inducing better environmental performance. The book contains 10 recommendations for improvement.

Evaluation: Good information on environmental performance indicators including analysis by sectors.

Application of Technology and Methodologies

Green Development, Integrating Ecology and Real Estate, **Alex Wilson, Jenifer L. Uncapher, Lisa McManigal, L. Hunter Lovins, Maureen Cureton, and William D. Browning, The Rocky Mountain Institute, John Wiley and Sons, New York, 1998.**

Essential how-to book for green development written for developers, architects, planners, engineers, and city officials. The authors found that in successful green developments, the development process was approached very differently than in conventional developments.

Evaluation: The book walks the reader through the green development process. Lots of case studies and tools.

The Industrial Green Game: Implications for Environmental Design and Management, **National Academy of Engineering, Deanna J. Richards, ed., National Academy Press, Washington, DC, 1997.**

This book discusses the applications of the concepts of industrial ecology to industrial practices. The first section is interesting but highly philosophical. The second section has a number of case studies regarding the implementation of industrial ecology principles. The third section has case examples.

Evaluation: A good, high-level discussion of industrial ecology. Unfortunately, the "green game" title tends to trivialize the implementation of industrial ecology principles.

Environmental Marketing, **Walter Coddington, McGraw Hill, Inc., New York, 1993.**

A marketing book focusing on selling environmentalism as a "new" consumer need and want.

Evaluation: The marketing department meets environmentalism and total quality management. Interesting from the standpoint that it was published at a time when environmental concerns were just reaching the consumer.

Greener Purchasing: Opportunities and Innovations, **by Trevor Russel, ed., Greenleaf Publishing, Sheffield, UK, 1998.**

This book contains a large collection of stories, case studies, innovations, and techniques involving green purchasing. Not only does it examine the issues, but it also provides a how-to guide for developing supply chain purchasing policies.

Evaluation: Comprehensive discussion on how public and private sector organizations are starting to compel their suppliers to deliver green products and services.

Sustainable Marketing—Managerial-Ecological Issues, **Donald A. Fuller, Sage Publications, Inc., Thousand Oaks, CA, 1999.**

The author discusses not only how to market green products and services, but also what products and services to produce in light of the concern over sustainability. The book offers a new paradigm that recognizes the limits of resources and ecosystems as well as a framework for sustainable marketing management. Chapters cover the design of sustainable products, sustainable marketing channels and communications, pricing, and market development.

Evaluation: An excellent discussion on how to apply sustainable development principles to marketing of the corresponding new products and services. Could be considered the next generation of Coddington's book Environmental Marketing.

***Streamlined Life-Cycle Assessment*, Thomas E Graedel, Prentice-Hall, Inc., Englewood Cliffs, NJ, 1998.**

This book examines the foundations and approaches to life-cycle assessment. It begins with the fundamental concerns about sustainable development and develops the basic concept and framework of life-cycle assessment. The book contains many technical details and is loaded with tools, illustrations, and case studies. Chapters 9 and 10 discuss life-cycle assessment as it pertains to facilities and services.

Evaluation: A basic reference and textbook for anyone involved in conducting life-cycle assessments.

***Sustainable Architecture: White Papers*, Earth Pledge Foundation, New York, 2000.**

A collection of short papers on various aspects of sustainable design and architecture.

Evaluation: Good source of ideas and information about what various groups are doing with regard to sustainable design.

***Strategy for Sustainable Business: Environmental Opportunity and Strategic Choice*, Liz Crosbie and Ken Knight, McGraw Hill Book Company, London, 1995.**

An early work describing how to apply sustainability principles to business. The book has four parts: (1) the context for sustainability in business, (2) the foundations for a corporate sustainability strategy, (3) the functional strategies of sustainability management, and (4) strategy implementation.

Evaluation: Although almost 10 years old, this well-organized book is a solid how-to guide for a company attempting to incorporate sustainable development into its strategies and operations.

***Deep Design: Pathways to a Livable Future*, David Wann, Island Press, Washington, DC, 1996.**

An early work that notes how our world, as currently designed, is flawed. The author argues that it makes better sense to design with nature instead of against it. The author describes this approach as Aikido Engineering. The approach appears to be a source of inspiration for some of William McDonough's criticisms of design.

Evaluation: Interesting stuff from a historical perspective. However, the information contained in the book has been addressed elsewhere in a more complete fashion.

***Sustainable Solutions: Developing Products and Services for the Future*, Martin Charter and Ursula Tischner, eds., Greenleaf Publishing, Sheffield, UK, February 2001.**

A very comprehensive book describing how companies are working to develop new products and services that adhere to the precepts of sustainability. The book covers the broad issues of sustainability, followed by methodologies for developing sustainable products and services. The last section contains case studies on the application of sustainable development principles.

Evaluation: A large book with a number of contributors. Plenty of information, models, and tools. Some of the chapters are reprints of notable journal articles.

***Mapping the Journey—Case Studies in Strategy and Action toward Sustainable Development*, Lorinda R. Rowledge, Russell S. Barton, and Kevin S. Brady, Greenleaf Publishing, Sheffield, UK, 1999.**

Case studies of companies and their journeys toward sustainable development. The work contains highlights of corporate strategies and approaches.

Evaluation: Many in-depth case studies that demonstrate how companies perceive the business case for sustainable development.

***Eco-efficiency: The Business Link to Sustainable Development*, Livio DeSimone, Frank Popoff, and the World Business Council for Sustainable Development, The MIT Press, Cambridge, MA, 1997.**

This seminal work lays out the concept of eco-efficiency, business's response to sustainable development and the 1992 Earth Summit. Eco-efficiency has five core themes: (1) an emphasis on service, (2) a focus on needs and quality of life, (3) consideration of the entire product life cycle, (4) recognition of the limits of ecocapacity, and (5) a process view—eco-efficiency is as much a journey as a destination. Eco-efficiency also has seven principles or goals that companies need to take into account to improve their operations, products, and services. These can be converted into measurable objectives: (1) reduce the material intensity of goods and services, (2) reduce the energy intensity of goods and services, (3) reduce toxic dispersion, (4) enhance material recyclability, (5) maximize sustainable use of renewable resources, (6) extend product durability, and (7) increase service intensity of goods and services.

Evaluation: A fine presentation of the eco-efficiency concept, written by business people for business people.

***Cannibals with Forks: The Triple Bottom Line of 21st Century Business*, John Elkington, New Society Publishers, Gabriola Island, BC, Canada, 1998.**

This book brings into question whether our capitalist economies are truly sustainable in their current form. He then poses a new, seven-dimension paradigm for creating a sustainable economy and future. Elkington is credited with inventing the term triple bottom line, a way of converting the principles of sustainable development into terms that business people can grasp. He extends the model into a geology metaphor, noting that the three bottom line elements—economic, environmental, and social—are in constant flux, shifting and rubbing together like tectonic plates. Each "shear zone" represents a set of issues that shift due to the pressures of social, political, economic, and technological development.

Evaluation: Interesting book by a well-respected author. While the Triple Bottom Line concept has taken hold, the tectonic plates have not, nor has his seven-dimension paradigm.

***Green Bottom Line: Environmental Accounting for Management Current Practice and Future Trends*, Martin Bennett and Peter James, eds., Greenleaf Publishing, Sheffield, UK, 1998.**

A sourcebook on environmental accounting. It brings together concepts, best practices, tools and case studies from many experts from both government and industry.

Evaluation: A little dated, but a good reference.

***Industrial Ecology: Policy Framework and Implementation*, Braden R. Allenby, Prentice Hall, Upper Saddle River, NJ, 1999.**

This book provides the concept and intellectual framework for industrial ecology, as Allenby states it: "the multidisciplinary study of industrial systems and their economic activities, and their links to fundamental natural systems." Allenby, the acknowledged guru of industrial ecology, begins with a definition and rationale for sustainable development. Using that platform, he lays out the principles of industrial ecology and discusses the various models and applications. Part 2 of the book contains a discussion on the legal and policy issues. Part 3 contains case studies.

Evaluation: Essentially the bible on industrial ecology.

***Biomimicry: Innovation Inspired by Nature*, Janine Benyus, William Morrow and Company, New York, NY, 1998.**

The author shows how innovations derived from nature and its processes can be captured and applied to business. Chapter 7 applies these ideas to business. This could be considered a companion book to Natural Capitalism and Cradle to Cradle.

Evaluation: An important book, easy to read.

***Ecological Design*, Sim Van Der Ryn and Stuart Cowan, Island Press, Washington, DC, 1996.**

The book expands upon the notion posed by Paul Hawken and William McDonough that our current problems of nonsustainability are the result of faulty design. The authors define "dumb design" as the mindless application by city planners, engineers, and other design professionals of standard design templates to buildings, facilities and infrastructure without regard to the needs of human communities and ecosystems. The book seeks to remedy this situation by offering three strategies for ecological design: conservation, regeneration, and stewardship. It also offers five broad principles of ecological design: (1) solutions grow from place, (2) ecological accounting informs design, (3) design with nature, (4) everyone is a designer, and (5) make nature visible.

Evaluation: A wake-up call for engineers and designers. Good starting point for anyone looking into creating a more sustainable design.

***Cradle to Cradle: Remaking the Way We Make Things*, William McDonough and Michael Braungart, North Point Press, New York, 2002.**

The authors' manifesto of eco-effectiveness, the sustainable development concept where all products and services are fashioned in a system that operates according to the rules of nature: nutrients and metabolisms. The authors offer a system composed of two cycles: biological and technical. Biological nutrients are consumed and returned to the biological cycle, where they are returned to nature as a biological nutrient. Technical nutrients are kept in the technical cycle to be used again and again.

Evaluation: A manifesto for a radically different philosophy and practice of manufacture and environmentalism. Lots of good ideas for engineering and design. However, the book takes issue with the concept of eco-efficiency, a favorite target of McDonough, saying that it is a philosophy for industry to just be less bad. In actuality, eco-effectiveness is the logical but idealistic endpoint for eco-efficiency.

***Designing Sustainable Communities—Learning from Village Homes*, Judy Corbett and Michael Corbett, Island Press, Washington, DC, 2000.**

A knowledgeable book about designing and building communities that meet the definition of sustainable. Discusses the history of sustainable community development.

Evaluation: A must-read for anyone engaged in sustainable community work.

The Case against Sustainable Development

Climate of Fear—Why We Shouldn't Worry about Global Warming, **Thomas Gale Moore, Cato Institute, Washington, DC, 1998.**

Thomas Gale Moore delivers the "so what" argument for global warming: that the predictions of global warming due to human activity may not be true, but even if they are a warmer climate will be better for the world, well, at least for the United States.

Evaluation: Well written and researched but with a clear U.S.-centered view of the issue. Important part of the counterliterature on environmental issues.

Earth Report 2000, **Ronald Bailey, ed., McGraw-Hill, New York, 2000.**

A collection of essays seeking to counter what the editor calls "the litany of gloomy predictions" about the decline of the environment.

Evaluation: A counterview to Lester Brown's State of the World publications.

Eco-Scam—The False Prophets of Ecological Apocalypse, **Ronald Bailey, St. Martin's Press, New York, 1993.**

An early work by Ronald Bailey that purports to debunk the myths about impending environmental crises offered by the doomsayer-experts. According to Bailey, following the Precautionary Principle is a bad idea.

Evaluation: Well-written critique but more of the same don't worry, be happy, technology will save the day logic.

The Skeptical Environmentalist: Measuring the Real State of the World, **Björn Lomborg, Cambridge University Press, Cambridge, UK, 2001.**

The author systematically examines the breadth of environmental issues and concludes that things are not as bad as many of the environmental alarmists make them out to be.

Evaluation: To me, Björn Lomborg represents the "don't worry, be happy, technology will save the day" crowd who has decided to take on the environmental "apocalyptics" like Paul Ehrlich and Lester Brown and throws in poor, long-dead Thomas Malthus for good measure. What gives him credibility is that he claims to be a scientist and former apocalyptic who, after some serious investigation, has seen the error of his ways. Björn Lomborg's assumes our world to be some huge, homogenous network of producers and consumers, all operating rationally in a frictionless global economy. In this world, all goods and services move about freely; technology is readily shared; the past is prologue; and we know everything there is to know about the earth, its ecosystems, and chemical and biological risks.

Hard Green—Saving the Environment from the Environmentalists, **Peter Huber, Basic Books, New York, 1999.**

Billed as the conservative manifesto and the answer to former vice president Al Gore's Earth in Balance, libertarian Peter Huber argues that those he calls the "soft greens" are actually doing more harm than good to the environment. He postulates that "hard green" is better: (1) technology and markets will handle all scarcity; (2) markets will handle pollution too, and the pollution we know about is all that counts; (3) hard technology is good, soft technology is not; and (4) wealth not frugality will save the environment.

Evaluation: Another, but more articulate, version of the "don't worry, be happy, technology will save the day" philosophy. However, this book is valuable in the sense that it debunks the apocalyptic view of how to address the world's environmental problems.

Sustainable Development Resources and Tools (On CD-ROM)

Current as of January 22, 2005. Readers are encouraged to check the external links and update as necessary. Broken links are often the result of web site reorganizations.

Item	Descriptions, excerpts from the Web sites	Link
Chapter 1: Introduction		
Text	Book text	
Links to documents		
Agenda 21	Report from the Earth Summit in Rio de Janeiro, 1992	http://habitat.igc.org/agenda21/
Our Common Future	The Brundtland Commission report (summary).	http://www.provincia.fe.it/agenda21/documenti/bruntland_report.htm
Johannesburg Summit 2002: Global Challenge, Global Opportunity	Trends in sustainable development. Statistics on trends and emerging global issues in sustainable development.	http://www.un.org/esa/sustdev/publications/critical_trends_report_2002.pdf
Rio Declaration on Sustainable Development	Principles from the United Nations Conference on Environment and Development, Rio de Janeiro, June 3–14, 1992.	http://habitat.igc.org/agenda21/rio-dec.htm
Sustainable development timeline	Salient events in the emergence of sustainable development. Poster-sized document made available through Environment Canada.	http://www.sdgateway.net/introsd/timeline.htm
Worldatch Environmental timeline	Traces key moments in the modern environmental movement from the 1960s to the present.	http://www.worldwatch.org/features/timeline/
Links to sustainable development-related organizations		
Business organizations		
The Brandt Forum	Sustainable development organization started by former West German chancellor Willy Brandt. Credited with first coining the term sustainable development seven years prior to the Brundtland Commission report.	http://brandt21forum.info/index.htm
Business for Social Responsibility (BSR)	Started in 1992, BSR provides tools, education, and advice to companies, helping them incorporate socially responsible business practices into their business operations.	www.bsr.org
Global Environmental Management Initiative (GEMI)	GEMI is a nonprofit organization that assists companies in fostering environmental, health, and safety excellence (EH&S) and corporate citizenship worldwide. Members share tools and information on EH&S matters.	www.gemi.org
World Business Council for Sustainable Development (WBCSD)	A coalition of 160 international companies from 30 countries and 20 major industrial sectors. WBCSD is recognized as the global business voice for sustainable development. It also has a global network of 40 national and regional business councils and partnerships involving approximately 1,000 organizations.	www.wbcsd.org

Public sector organizations

Air Force Center for Environmental Excellence (AFCEE) Sustainable Development	AFCEE/EQ teams with AFCEE's Design and Construction Directorate to provide technical resources to support air force units on their journey to sustainability.	www.afcee.brooks.af.mil/ eq/programs/progpage. asp?PID=27
Department of Energy, Energy Efficiency and Renewable Energy	Part of the energy efficiency and renewable energy network. Has useful information on the concept of sustainable development, including overview articles, slide shows, links to other sources of information, recommended books and videos, and educational materials and programs.	http://www.eere.energy. gov/
DOE Office of Energy Efficiency and Renewable Energy (EERE)	Helps home and business owners reduce their use of natural gas and electricity. (Benefits of conservation include environmental benefits and consumer protection from rising energy costs.)	http://www.eere.energy. gov/
International Council for Local Environmental Initiatives (ICLEI)	ICLEI is an international association of local governments implementing sustainable development. It operates as an international environmental agency for local governments.	http://www.iclei.org
National Renewable Energy Laboratory (NREL)	The National Renewable Energy Laboratory (NREL) conducts research on renewable energy and energy efficiency.	http://www.nrel.gov/
United Nations Division for Sustainable Development	The Division for Sustainable Development serves as the substantive secretariat responsible for servicing the Commission on Sustainable Development; and for follow-up of the implementation of Agenda 21, as well as the Plan of Implementation (POI) of the 2002 World Summit on Sustainable Development.	www.un.org/esa/sustdev
U.S. Business Council for Sustainable Development (US BCSD)	A regional partner organization and the U.S. business voice of sustainable development to the WBCSD. Delivers projects demonstrating the value of sustainability to business.	http://www.usbcsd.org/

Nongovernmental organizations

Coalition for Environmentally Responsible Economies (CERES)	A U.S. coalition of environmental, investor, and advocacy groups working together for a sustainable future. Seventy companies have endorsed the CERES Principles, a 10-point code of environmental conduct.	http://www.ceres.org/
Forum for the Future	A United Kingdom organization doing projects and programs to accelerate the movement toward sustainability.	http://www.forumforthefu- ture.org.uk/default.asp
Global Reporting Initiative (GRI)	The GRI is a multistakeholder process and institution whose mission is to develop and disseminate its Sustainability Reporting Guidelines.	www.globalreporting.org
Green Seal	Evaluates products and provides credible and objective information to direct the purchaser to environmentally responsible products and services.	http://www.greenseal. org/index.html

International Institute for Sustainable Development (IISD)	IISD's mission is to champion innovation, enabling societies to live sustainably. The organization receives support from the government of Canada, plus project funding from the government of Canada, the province of Manitoba, other national governments, United Nations agencies, foundations, and the private sector.	www.iisd.org
Millennium Ecosystem Assessment (MA)	The MA is an international work program designed to assist decision makers and the public by providing scientific information concerning the consequences of ecosystem change for human well-being and options for responding to those changes.	http://www.millenniumas-sessment.org/en/index.aspx
National Round Table on the Environment and the Economy (NRTEE)	NRTEE is an independent advisory body that provides decision makers, opinion leaders, and the Canadian public with advice and recommendations for promoting sustainable development.	www.nrtee-trnee.ca
The Natural Step	The Natural Step, founded in 1989, is a nonprofit advisory organization and think tank that fosters the application of The Natural Step principles.	www.naturalstep.org
Population Reference Bureau	The Population Reference Bureau provides information about the population dimensions of important social, economic, and political issues. Its mission is to be the leader in providing timely and objective information on U.S. and international population trends and their implications.	http://www.prb.org/
Smart Growth Network	A network of environmental groups, historic preservation organizations, professional organizations, developers, real estate interests, and local and state government entities that encourage sustainable develop-ment.	http://www.smartgrowth.org/
Sprawl City	Helps people understand the effects of urban sprawl and allows the public to make better use of available federal data.	http://www.sprawlcity.org/
Sustainable Buildings Industry Council (SBIC)	A nonprofit organization to advance the design, affordability, energy performance, and environmental soundness of residential, institutional, and commercial buildings nationwide.	http://www.psic.org/
Resources for the Future (RFF)	A respected think tank that analyzes environ-mental, energy, and natural resource topics.	http://www.rff.org/
Rocky Mountain Institute (RMI)	A nonprofit institute established in 1982 by resource analysts L. Hunter Lovins and Amory B. Lovins. It fosters the efficient and restorative use of natural, human, and other capital.	http://www.rmi.org/
Trust for Public Land	A national nonprofit working to protect land for human enjoyment and well-being and to improve the health and quality of life of American communities.	http://www.tpl.org/index.cfm

U.S. Green Building Council (USGBC)	The USGBC is managing a national consensus process for producing high-performance buildings. Council members work together to develop LEED products and resources, policy guidance, and educational and marketing tools that support the adoption of sustainable building principles.	www.usgbc.org
World Resources Institute	An independent nonprofit organization focused on achieving progress on four goals: (1) protect the earth's living systems, (2) increase access to information, (3) create sustainable enterprise and opportunity, and (4) reverse global warming.	www.wri.org/index.html
Worldwatch Institute	Founded by Lester Brown in 1974, the Worldwatch Institute is a source of information on key environmental, social, and economic trends.	www.worldwatch.org

Professional societies and trade associations

American Society of Civil Engineers (ASCE)	A professional organization representing more than 130,000 civil engineers. Has published a policy on sustainable development for civil engineers.	www.asce.org
American Council of Engineering Companies (ACEC)	An engineering trade organization numbering more than 5,800 firms throughout the country, engaged in a wide range of engineering works.	www.acec.org
American Institute of Chemical Engineers (AIChE), Institute for Sustainability (IfS)	The purpose of the AIChE IfS is to promote awareness of the scientific and engineering challenges of sustainability, develop practices and tools that will guide the creation of more sustainable manufacturing processes, and support and encourage chemical engineers' contributions toward meeting the needs of tomorrow's world.	www.aiche.org/sustainability
American Society of Mechanical Engineers (ASME)	Founded in 1880 as the American Society of Mechanical Engineers, today's ASME is a 120,000-member professional organization focused on technical, educational, and research issues of the engineering and technology community.	http://www.asme.org/
Institute of Electrical and Electronics Engineers (IEEE)	IEEE is a technical professional association of more than 360,000 individual members in approximately 175 countries. Through its members, IEEE is an authority in technical areas ranging from computer engineering, biomedical technology, and telecommunications, to electric power, aerospace, and consumer electronics, among others.	http://ieee.org/portal/index.jsp
International Federation of Consulting Engineers (FIDIC)	FIDIC (the acronym represents the French version of the name) represents the international business interests of firms belonging to national member associations of engineering-based consulting companies. The member firms of each national association comply with FIDIC's Code of Ethics, Policy Statements, and Statutes.	www.fidic.org

Academic organizations		
Engineers Without Borders-USA	Engineers Without Borders–USA (EWB-USA) is a nonprofit organization established in 2000 to help developing areas worldwide with their engineering needs, while involving and training a new kind of internationally responsible engineering student. EWB-USA projects involve the design and construction of water, wastewater, sanitation, energy, and shelter systems.	www.ewb-usa.org
WFEO ComTech	WFEO was established with the support of UNESCO in 1968. WFEO ComTech is the standing committee whose mission is to lead the engineering profession world-wide in the promotion and application of sustainable technology and, accordingly, disseminate information about sustainable technologies.	www.wfeo-comtech.org
Related software		
Dashboard of Sustainability	The Dashboard of Sustainability is a free, noncommercial software that presents the complex relationships between economic, social, and environmental issues in a highly communicative format aimed at decision makers and citizens interested in sustainable development.	http://esl.jrc.it/envind/dash-brds.htm

Chapter 2: Sustainable Development: Origins, Concepts, and Principles

Text	Book text	
Links to publications		
Sustainable development business toolkit	The International Chamber of Commerce (ICC) has developed this tool kit, which contains some of the latest thinking on several key issues and could be useful as an organization seeks to improve its environmental performance and sustainability.	http://www.iccwbo.org/home/environment_and_energy/sdcharter/services/toolkit/toolkit.asp
Links to sustainable development principles		
Sustainable development principles (IISD)	The principles included in this site contain elements that address the three major aspects of sustainable development (environment, economy, and community). Principles are linked to an authorized, full text version.	http://www.iisd.org/sd/principle.asp
Smart Communities Network	List of sustainable development principles provided by the Smart Communities Network.	http://www.sustainable.doe.gov/overview/principles.shtml
Additional sustainable development concepts		
Design for Environment (DfE)	The DfE program is one of the Environmental Protection Agency's (EPA's) premier partnership programs, working with individual industry sectors to compare and improve the performance and human health and environmental risks and costs of existing and alternative products, processes, and practices. DfE partnership projects promote integrating cleaner, cheaper, and smarter solutions into everyday business practices.	http://www.epa.gov/opptintr/dfe/index.htm
Five Capitals model	From the Forum for the Future Web site. The model describes five types of sustainable capital from which we derive the goods and services we need to improve the quality of our lives.	http://www.forumforthefuture.org.uk/aboutus/fivecapitalsmodel_page814.aspx
Industrial Ecology	Industrial Ecology aims to incorporate the cyclical patterns of ecosystems into designs for industrial production processes that will work in unison with natural systems.	http://www.sustainable.doe.gov/business/parkintro.shtml
Triple Bottom Line (TBL)	The TBL focuses corporations not just on the economic value they add, but also on the environmental and social value they add—and destroy. At its narrowest, the term triple bottom line is used as a framework for measuring and reporting corporate performance against economic, social, and environmental parameters.	http://www.sustainability.com/philosophy/triple-bottom/tbl-intro.asp
Organizations		
Business and Sustainable Development	This Web site explains the strategies and tools that companies can draw on to translate an aspiration of sustainability into practical, effective solutions. Case studies from around the world are provided as an example of each measure.	http://www.bsdglobal.com/

Business for Social Responsibility (BSR)	Since 1992, BSR has helped companies of all sizes and sectors to achieve success in ways that demonstrate respect for ethical values, people, communities, and the environment.	http://www.bsr.org/
Ethical Corporation	An independent business information provider and events producer on the issues in and around corporate social, financial, and environmental responsibility.	http://www.ethicalcorp.com/

Related software

Buried Treasure Online	SustainAbility's Web site in which it presents the Sustainable Business Value Matrix.	http://www.sustainability.com/business-case/contents.asp
Green and Gold	Provides access to a CD-ROM: Sustainable Development Principles in Action: Learning from the Sydney 2000 Experience.	http://www.greengold.on.ca/publications/sydneycd.html

Chapter 3: Sustainable Development: Client Needs and Market Drivers

Text	Book text	

Links to documents

Tomorrow's Markets	This publication joins the World Business Council on Sustainable Development, the World Resources Institute, and the United Nations Environment Programme to identify the trends that are shaping the global business environment. These trends are shaping a new marketscape, the landscape through which business must navigate to succeed.	http://business.wri.org/pubs_content.cfm?PubID=3155
World Resources 2002–2004	Tenth in the biennial World Resources series on the global environment, the report defines governance in everyday terms, with reference to a wealth of case studies. It assesses the state of environmental governance in nations around the world and summarizes results from the Access Initiative, a first-ever attempt to systematically measure governments' performance in providing their citizens access to environmental information, decision making, and justice.	http://governance.wri.org/pubs_description.cfm?PubID=3764

Links to related organizations

Convention on Biological Diversity	Part of the United Nations Environment Programme. Information on biodiversity.	http://www.biodiv.org/default.aspx

Chapter 4: Walking the Talk

Text	Book text	
Links to documents		
SIGMA project	The SIGMA project aims to provide clear, practical advice to organizations to help them make a meaningful contribution to sustainable development.	http://www.projectsigma. com/Default.asp
Related software		
Global Environmental Management Initiative (GEMI)	Various software tools for environmental management and sustainability, including SD Planner.	http://www.green-biz.com/frame/1. cfm?targetsite=http://www. gemi.org

Chapter 5: Greening the Engineering Company

Text	Book text	
Environmental management systems		
Global Environmental Management Initiative (GEMI) HSE Web Depot	The GEMI HSE (Health, Safety and Environment) Web Depot is a Web-based information resource for HSE management information systems. The HSE Web Depot presents a framework for HSE-MIS planning, development, and system rollout and improvement. It organizes company experiences within these areas into an easy-to-use format.	http://www.hsewebdepot. org
ISO 14000 / ISO 14001 Environmental Management Guide	ISO 14000, ISO14001, ISO 14004 ... the myriad ISO 14000 standards and information related to environmental management can sometimes hinder progress and cause confusion. This Web site is designed to untangle and simplify these standards to make environmental management using the standards a much easier task.	http://www.iso14000-iso14001-environmental-management.com/
ISO 14001 "Pizza"	A different way of looking at ISO 14001 standards.	http://www.mgmt14k. com/014kpizza.htm
U.S. Environmental Protection Agency Environmental Management Systems	This Web site provides information and resources related to Environmental Management Systems (EMSs) for businesses, associations, the public, and state and federal agencies. An EMS is a set of processes and practices that enable an organization to reduce its environmental impacts and increase its operating efficiency.	http://www.epa.gov/ems/
Performance rating systems		
PROPER	Program for Pollution Control Evaluation and Rating. A Model for Promoting Environmental Compliance and Strengthening Transparency and Community Participation in Developing Countries. Source: Shakeb Afsah International Resources Group Ltd., Washington, DC.	http://www.worldbank.org/ nipr/work_paper/PROPER2. pdf
Ecological Footprint	Online calculator. How big is your personal ecological footprint?	http://www.myfootprint.org/
Information resources		
Cleaner Technologies Substitutes Assessment	U.S. EPA's *Cleaner Technologies Substitutes Assessment —A Methodology and Resource Guide.*	http://www.epa.gov/dfe/ pubs/tools/ctsa/

Energy Star	Energy Star is a government-backed program helping businesses and individuals protect the environment through superior energy efficiency.	http://www.energystar.gov/
Estimating Pollution Load: The Industrial Pollution Projection System (IPPS)	IPPS is a modeling system that combines data from industrial activity (such as production and employment) with data on pollution emissions to calculate *pollution intensity* factors, i.e., the level of pollution emissions per unit of industrial activity.	http://www.worldbank.org/nipr/ipps/ippsweb.htm
ReThink Paper	Strategies to reduce paper usage and encourage paper recycling.	http://www.rethinkpaper.org/
WasteWise Resource Management: Innovative Solid Waste Contracting Methods	To help organizations gain a firm grasp on the concept of resource management and negotiate their waste-hauling contracts to focus on resource conservation, WasteWise developed this extensive how-to manual (available for downloading).	http://www.epa.gov/wastewise/wrr/rm.htm#manual

Sustainable development reporting

AA1000 Framework	The framework is designed to improve accountability and performance by learning through stakeholder engagement. It has been used worldwide by leading businesses, nonprofit organizations, and public bodies. The framework helps users to establish a systematic stakeholder engagement process that generates the indicators, targets, and reporting systems needed to ensure its effectiveness in overall organizational performance.	http://www.accountability.org.uk/aa1000/default.asp
U.S. Environmental Protection Agency Climate Leaders	Summary of Climate Leaders GHG Inventory Guidance. The Climate Leaders greenhouse gas (GHG) inventory guidance defines how partners inventory and report their GHG emissions.	http://www.epa.gov/climate-leaders/protocol.html
Deloitte Sustainability Reporting Scorecard	Scorecard used in rating a corporate sustainability report.	http://www.deloitte.com/dtt/cda/doc/content/FullScorecardpdf(3).pdf
Department for Environment Food and Rural Affairs (Defra)	In October 2000, the prime minister challenged the top 350 UK companies to produce environmental reports by the end of 2001. Defra has produced a general set of guidelines that set out in straightforward terms how to produce a good-quality environmental report.	http://www.defra.gov.uk/environment/envrp/general/pdf/envrptgen.pdf
GreenBiz.com	Corporate reporting: the big picture. Approaches for preparing a sustainability report.	http://www.greenbiz.com/toolbox/essentials_third.cfm?LinkAdvID=25026
Sustainability Reporting Guidelines (Global Reporting Initiative)	The GRI guidelines represent the foundation upon which all other GRI reporting documents are based and outline core content that is broadly relevant to all organizations regardless of size, sector, or location.	http://www.globalreporting.org/guidelines/2002.asp
World Business Council for Sustainable Development reporting portal	The reporting portal provides a picture of issues companies are currently tackling in their sustainable development reports and the kind of information they are presenting.	http://www.sdportal.org/templates/Template3/layout.asp?MenuId=38

Related software		
GHG Calculation Tools	Tools produced by the World Business Council for Sustainable Development (WBCSD). They have been peer-reviewed and tested by experts and industry leaders and represent a "best practice" for emission calculation tools.	http://www.ghgprotocol.org/standard/tools.htm
Links to documents		
Greening Industry	Publication from the World Bank discussing a new model for creating a more environmentally benign industrial community.	http://www.worldbank.org/nipr/greening/full_text/index.htm
Links to related organizations		
Canadian Department of Foreign Affairs and International Trade (DFAIT)	Information about greening programs as applied to the Canadian government's operations.	http://www.dfait-maeci.gc.ca/sustain/EnvironMan/system/greenop/GreenHQ/GreenHQ-en.asp
Green Seal	Green Seal is an independent, nonprofit organization that strives to achieve a healthier and cleaner environment by identifying and promoting products and services that cause less toxic pollution and waste, conserve resources and habitats, and minimize global warming and ozone depletion.	http://www.greenseal.org/index.html
GreenBiz.com Business Toolbox	Ideas on making business operations more green.	http://www.greenbiz.com/toolbox/
Greening Federal Facilities	*Greening Federal Facilities*, second edition, is a nuts-and-bolts resource guide designed to increase energy and resource efficiency, cut waste, and improve the performance of federal buildings and facilities.	http://www.eere.energy.gov/femp/pdfs/29267-0.pdf
Greening the National Park Service	Greening the National Park Service. The "Green Toolbox"	http://www.nps.gov/renew/toolbox.htm
Related software		
FinePrint software	FinePrint is a Windows printer driver that saves ink, paper, and time by controlling and enhancing printed output.	http://www.fineprint.com/products/fineprint/index.html

Chapter 6: Delivering a Sustainable Project

Text	Book text	
Links to documents		
The Guidelines for Sustainable Buildings (Stanford University)	The guidelines provide an overview and introduction for designing sustainable buildings. It explains how the word *sustainability* is used in the context of this document and why it is important at Stanford University. (Section 2, "Process Phases," describes the process for implementing sustainable principles in a building project, with a discussion of sustainability issues for each phase of design and construction. The technical guidelines for sustainability are contained in Section 3.)	http://cpm.stanford.edu/process_new/Sustainable_Guidelines.pdf

High Performance Building Guidelines	Guidelines from the City of New York, Department of Building and Construction. High-performance buildings maximize operational energy savings; improve comfort, health, and safety of occupants and visitors; and limit detrimental effects on the environment.	http://www.nyc.gov/html/ddc/html/ddcgreen/highperf.html
LEED Green Building Rating System	Developed by the USGBC membership, the Leadership in Energy and Environmental Design (LEED) Green Building Rating System is a national, consensus-based, market-driven building rating system designed to accelerate the development and implementation of green building practices.	http://www.usgbc.org/AboutUs/programs.asp
Minnesota Sustainable Design Guide	*The Minnesota Sustainable Design Guide* educates and assists architects, building owners, occupants, educators, students, and the general public concerning sustainable building design. The guide is a design tool that can be used to overlay environmental issues on the design, construction, and operation of both new and renovated facilities.	http://www.sustainabledesignguide.umn.edu/

Links to related organizations

BetterBricks	BetterBricks is dedicated to raising awareness and demand for energy efficiency in buildings by sharing information and resources with the people who design, own, and operate them.	http://www.betterbricks.com/default.aspx
BRE's Environmental Assessment Method (BREEAM)	BREEAM has been used to assess the environmental performance of both new and existing buildings.	http://products.bre.co.uk/breeam/index.html
International Initiative for Sustainable Built Environment (iiSBE)	IiSBE is an international nonprofit organization whose overall aim is to actively facilitate and promote the adoption of policies, methods, and tools to accelerate the movement toward a global sustainable built environment.	http://greenbuilding.ca/index.html
Sustainable Building and Construction Forum	The Sustainable Building and Construction Forum was set up to facilitate dialogue and exchange of information among key stakeholders, and with their constituencies, on issues related to sustainability in building and construction.	http://www.unep.or.jp/ietc/sbc/index.asp

Related software

Eco-it (demonstration software)	Eco-it software allows the user to model a complex product and its life cycle in a few minutes. Eco-it immediately calculates the environmental load and shows which parts of the product contribute most.	http://www.pre.nl/eco-it/default.htm
SimaPro (demonstration software)	A professional software tool to collect, analyze, and monitor the environmental performance of products and services. Allows the user to model and analyze complex life cycles in a systematic and transparent way, following the ISO 14040 series recommendations.	http://www.pre.nl/simapro/default.htm

Chapter 7: The Future of Sustainable Development

Text	Book text	
Links to documents		
Can OECD Countries Put Theory into Practice? A Blueprint for Progress Toward Sustainable Development	Despite strong economic growth in the decade since the Earth Summit in 1992, progress toward eradicating poverty and addressing pressing environmental problems at the global level has remained limited. Organization for Economic Cooperation and Development (OECD) countries can turn the rhetoric about sustainable development into action through a range of concrete measures.	http://www.isuma.net/ v03n02/johnston/ johnston_e.shtml
Envisioning the Global Environment in 2025	General Motors began a multiyear cooperative envisioning project with the Greening of Industry Network (GIN) in 1998 to develop a shared understanding of the forces of global environment change in 2025. Twelve themes covering a wide variety of areas (i.e., mobility, business cooperation, human settlement, etc.) were rated as of major-to-critical importance to respondents' countries.	http://www.greeningofindustry.org/gm.html
GEO: Global Environment Outlook 3 *Past, Present and Future Perspectives*	GEO-3 provides an overview of the main environmental developments over the past three decades and how social, economic, and other factors have contributed to the changes that have occurred.	http://www.grida.no/geo/ geo3/index.htm
"Visions of sustainability in 2050"	Article by Anthony D. Cortese, ScD, president of Second Nature. It is a vision for a healthy, peaceful, socially just, economically secure and environmentally sustainable world.	http://www.secondnature. org/pdf/snwritings/articles/ 2050SustVision.pdf
Links to related organizations		
EcoFuture	Web site with links to other sustainable development-related sites.	http://www.ecofuture.org/

APPENDIX C

Sustainable Development Principles

Appendix C presents in detail several sets of sustainable development-related principles used to guide the activities of public and private organizations and of civil society. The following principles are discussed here. A longer list of principles can be found on the International Institute for Sustainable Development Web site at http://www.iisd.org/sd/principle.asp.

- The Earth Charter
- Ahwahnee Principles
- The Natural Step
- Caux Round Table Principles for Business
- CERES Principles
- Enlibra
- Hannover Principles
- Bellagio Principles

C.1. The Earth Charter[1]

The Earth Charter is a synthesis of values, principles, and aspirations that are shared by a growing number of women, men, and organizations around the world. Drafting the Earth Charter was part of the unfinished business of the Earth Summit. The Earth Charter was written with extensive international consultations conducted over many years. Currently, the Earth Charter is being disseminated to individuals and organizations in all sectors of society throughout the world, and it says, in part:

We urgently need a shared vision of basic values to provide an ethical foundation for the emerging world community. Therefore, together in hope we affirm the following interdependent principles for a sustainable way of life as a common standard by which the conduct of all individuals, organizations, businesses, governments, and transnational institutions is to be guided and assessed.

I. Respect and care for the community of life

> *1. Respect Earth and life in all its diversity.*

> *2. Care for the community of life with understanding, compassion, and love.*

> *3. Build democratic societies that are just, participatory, sustainable, and peaceable.*

> *4. Secure Earth's bounty and beauty for present and future generations.*

II. Ecological Integrity

> *1. Protect and restore the integrity of Earth's ecological systems, with special concern for biological diversity and the natural processes that sustain life.*

> *2. Prevent harm as the best method of environmental protection and, when knowledge is limited, apply a precautionary approach.*

> *3. Adopt patterns of production, consumption, and reproduction that safeguard Earth's regenerative capacities, human rights, and community well-being.*

> *4. Advance the study of ecological sustainability and promote the open exchange and wide application of the knowledge acquired.*

III. Social and Economic Justice

> *1. Eradicate poverty as an ethical, social, and environmental imperative.*

> *2. Ensure that economic activities and institutions at all levels promote human development in an equitable and sustainable manner.*

3. Affirm gender equality and equity as prerequisites to sustainable development and ensure universal access to education, healthcare, and economic opportunity.

4. Uphold the right of all, without discrimination, to a natural and social environment supportive of human dignity, bodily health, and spiritual well-being, with special attention to the rights of indigenous peoples and minorities.

IV. Democracy, Nonviolence, and Peace

1. Strengthen democratic institutions at all levels, and provide transparency and accountability in governance, inclusive participation in decision making, and access to justice.

2. Integrate into formal education and life-long learning the knowledge, values, and skills needed for a sustainable way of life.

3. Treat all living beings with respect and consideration. 4. Promote a culture of tolerance, nonviolence, and peace.

C.2. Ahwahnee Principles

The Ahwahnee Principles were delivered in the fall of 1991 to approximately 100 local officials at a conference in the Ahwahnee Hotel in Yosemite National Park, California. The principles were the result of efforts of Peter Katz, a staff member of the Local Government Commission and later the author of the book *The New Urbanism: Toward an Architecture of Community.* Katz brought together a number of leading architects to draft a vision of community architecture for local officials, an alternative to urban sprawl.

Preamble

Existing patterns of urban and suburban development seriously impair our quality of life. The symptoms are: more congestion and air pollution resulting from our increased dependence on automobiles, the loss of precious open space, the need for costly improvements to roads and public services, the inequitable distribution of economic resources, and the loss of a sense of community. By drawing upon the best from the past and the present, we can plan communities that will more successfully serve the needs of those who live and work within them. Such planning should adhere to certain fundamental principles.

Community Principles:

1. *All planning should be in the form of complete and integrated communities containing housing, shops, work places, schools, parks and civic facilities essential to the daily life of the residents.*

2. *Community size should be designed so that housing, jobs, daily needs and other activities are within easy walking distance of each other.*

3. *As many activities as possible should be located within easy walking distance of transit stops.*

4. *A community should contain a diversity of housing types to enable citizens from a wide range of economic levels and age groups to live within its boundaries.*

5. *Businesses within the community should provide a range of job types for the community's residents.*

6. *The location and character of the community should be consistent with a larger transit network.*

7. *The community should have a center focus that combines commercial, civic, cultural and recreational uses.*

8. The community should contain an ample supply of specialized open space in the form of squares, greens and parks whose frequent use is encouraged through placement and design.

9. Public spaces should be designed to encourage the attention and presence of people at all hours of the day and night.

10. Each community or cluster of communities should have a well-defined edge, such as agricultural greenbelts or wildlife corridors, permanently protected from development.

11. Streets, pedestrian paths and bike paths should contribute to a system of fully- connected and interesting routes to all destinations. Their design should encourage pedestrian and bicycle use by being small and spatially defined by buildings, trees and lighting; and by discouraging high speed traffic.

12. Wherever possible, the natural terrain, drainage and vegetation of the community should be preserved with superior examples contained within parks or greenbelts.

13. The community design should help conserve resources and minimize waste.

14. Communities should provide for the efficient use of water through the use of natural drainage, drought tolerant landscaping and recycling.

15. The street orientation, the placement of buildings and the use of shading should contribute to the energy efficiency of the community.

Regional Principles:

1. The regional land-use planning structure should be integrated within a larger transportation network built around transit rather than freeways.

2. Regions should be bounded by and provide a continuous system of greenbelt/wildlife corridors to be determined by natural conditions.

3. Regional institutions and services (government, stadiums, museums, etc.) should be located in the urban core.

4. Materials and methods of construction should be specific to the region, exhibiting a continuity of history and culture and compatibility with the climate to encourage the development of local character and community identity.

Implementation Principles:

1. The general plan should be updated to incorporate the above principles.

2. Rather than allowing developer-initiated, piecemeal development, local governments should take charge of the planning process. General plans should designate where new growth, infill or redevelopment will be allowed to occur.

3. Prior to any development, a specific plan should be prepared based on these planning principles.

4. Plans should be developed through an open process and participants in the process should be provided visual models of all planning proposals.

C.3. The Natural Step

1. *In order for a society to be sustainable, nature's functions and diversity are not systematically subject to increasing concentrations of substances extracted from the Earth's crust.*

   Materials such as minerals, metals and fossil fuels must not be extracted faster than they are re-deposited back into the earth. Once extracted, they must be used in such a way as not to harm the ecosystems. Products and services that use minerals, metals and other non-renewable materials should be designed so that they can be easily separated after use to be recycled and used again. Substitutes for non-renewable fuels must be found and utilized.

2. *In order for a society to be sustainable, nature's functions and diversity are not systematically subject to increasing concentrations of substances produced by society.*

   The compounds produced by society must not be deposited in the earth's biosphere at a rate faster than they can be degraded or safely assimilated.

3. *In order for a society to be sustainable, nature's functions and diversity are not systematically impoverished by displacement, over-harvesting, or other forms of ecosystem manipulation.*

   Nature's diversity and ecosystem services must not be degraded.

4. *Resources are used efficiently and fairly in order to meet basic human needs globally.*

   The earth's renewable and non-renewable resources and ecosystems must be used in a way that meets the basic needs of all people.

C.4. Caux Roundtable Principles for Business[2]

The Caux Round Table is an international network of business leaders working to promote moral capitalism. The group was formed in 1986 by Frederick Phillips, former president of Philips Electronics, and Olivier Giscard d'Estaing, former vice-chairman of INSEAD. The group recognized the important role of corporations in reducing threats to world peace and set about developing its Principles for Business through a series of dialogues among business leaders from Europe, Japan, and the United States. These principles were first published in 1994 and presented to the United Nations World Summit on Social Development in 1995.

Introduction

The Caux Round Table believes that the world business community should play an important role in improving economic and social conditions. As a statement of aspirations, this document aims to express a world standard against which business behavior can be measured. We seek to begin a process that identifies shared values, reconciles differing values, and thereby develops a shared perspective on business behavior acceptable to and honored by all.

These principles are rooted in two basic ethical ideals: kyosei and human dignity. The Japanese concept of kyosei means living and working together for the common good enabling cooperation and mutual prosperity to coexist with healthy and fair competition. "Human dignity" refers to the sacredness or value of each person as an end, not simply as a means to the fulfillment of others' purposes or even majority prescription.

The General Principles in Section 2 seek to clarify the spirit of kyosei and "human dignity," while the specific Stakeholder Principles in Section 3 are concerned with their practical application.

In its language and form, the document owes a substantial debt to The Minnesota Principles, a statement of business behavior developed by the Minnesota Center for Corporate Responsibility. The Center hosted and chaired the drafting committee, which included Japanese, European, and United States representatives.

Business behavior can affect relationships among nations and the prosperity and well-being of us all. Business is often the first contact between nations and, by the way in which it causes social and economic changes, has a significant impact on the level of fear or confidence felt by people worldwide. Members of the Caux Round Table place their first emphasis on putting one's own house in order, and on seeking to establish what is right rather than who is right.

Section 1. Preamble

The mobility of employment, capital, products and technology is making business increasingly global in its transactions and its effects.

Law and market forces are necessary but insufficient guides for conduct.

Responsibility for the policies and actions of business and respect for the dignity and interests of its stakeholders are fundamental.

Shared values, including a commitment to shared prosperity, are as important for a global community as for communities of smaller scale.

For these reasons, and because business can be a powerful agent of positive social change, we offer the following principles as a foundation for dialogue and action by business leaders in search of business responsibility. In so doing, we affirm the necessity for moral values in business decision making. Without them, stable business relationships and a sustainable world community are impossible.

Section 2. General Principles

Principle 1. The Responsibilities Of Businesses: Beyond Shareholders toward Stakeholders

The value of a business to society is the wealth and employment it creates and the marketable products and services it provides to consumers at a reasonable price commensurate with quality. To create such value, a business must maintain its own economic health and viability, but survival is not a sufficient goal. Businesses have a role to play in improving the lives of all their customers, employees, and shareholders by sharing with them the wealth they have created. Suppliers and competitors as well should expect businesses to honor their obligations in a spirit of honesty and fairness. As responsible citizens of the local, national, regional and global communities in which they operate, businesses share a part in shaping the future of those communities.

Principle 2. The Economic and Social Impact of Business: Toward Innovation, Justice and World Community

Businesses established in foreign countries to develop, produce or sell should also contribute to the social advancement of those countries by creating productive employment and helping to raise the purchasing power of their citizens. Businesses also should contribute to human rights, education, welfare, and vitalization of the countries in which they operate.

Businesses should contribute to economic and social development not only in the countries in which they operate, but also in the world community at large, through effective and prudent use of resources,

191

free and fair competition, and emphasis upon innovation in technology, production methods, marketing and communications.

Principle 3. Business Behavior: Beyond the Letter of Law Toward a Spirit of Trust

While accepting the legitimacy of trade secrets, businesses should recognize that sincerity, candor, truthfulness, the keeping of promises, and transparency contribute not only to their own credibility and stability but also to the smoothness and efficiency of business transactions, particularly on the international level.

Principle 4. Respect for Rules

To avoid trade frictions and to promote freer trade, equal conditions for competition, and fair and equitable treatment for all participants, businesses should respect international and domestic rules. In addition, they should recognize that some behavior, although legal, may still have adverse consequences.

Principle 5. Support for Multilateral Trade

Businesses should support the multilateral trade systems of the GATT/World Trade Organization and similar international agreements. They should cooperate in efforts to promote the progressive and judicious liberalization of trade and to relax those domestic measures that unreasonably hinder global commerce, while giving due respect to national policy objectives.

Principle 6. Respect for the Environment

A business should protect and, where possible, improve the environment, promote sustainable development, and prevent the wasteful use of natural resources.

Principle 7. Avoidance of Illicit Operations

A business should not participate in or condone bribery, money laundering, or other corrupt practices: indeed, it should seek cooperation with others to eliminate them. It should not trade in arms or other materials used for terrorist activities, drug traffic or other organized crime.

Section 3. Stakeholder Principles

Customers

We believe in treating all customers with dignity, irrespective of whether they purchase our products and services directly from us or otherwise acquire them in the market. We therefore have a responsibility to:

- *provide our customers with the highest quality products and services consistent with their requirements;*

- *treat our customers fairly in all aspects of our business transactions, including a high level of service and remedies for their dissatisfaction;*

- *make every effort to ensure that the health and safety of our customers, as well as the quality of their environment, will be sustained or enhanced by our products and services;*

- *assure respect for human dignity in products offered, marketing, and advertising; and respect the integrity of the culture of our customers.*

Employees

We believe in the dignity of every employee and in taking employee interests seriously. We therefore have a responsibility to:

- provide jobs and compensation that improve workers' living conditions;

- provide working conditions that respect each employee's health and dignity;

- be honest in communications with employees and open in sharing information, limited only by legal and competitive constraints;

- listen to and, where possible, act on employee suggestions, ideas, requests and complaints;

- engage in good faith negotiations when conflict arises;

- avoid discriminatory practices and guarantee equal treatment and opportunity in areas such as gender, age, race, and religion;

- promote in the business itself the employment of differently abled people in places of work where they can be genuinely useful;

- protect employees from avoidable injury and illness in the workplace;

- encourage and assist employees in developing relevant and transferable skills and knowledge; and

- be sensitive to the serious unemployment problems frequently associated with business decisions, and work with governments, employee groups, other agencies and each other in addressing these dislocations.

Owners/Investors

We believe in honoring the trust our investors place in us. We therefore have a responsibility to:

- apply professional and diligent management in order to secure a fair and competitive return on our owners' investment;

- disclose relevant information to owners/investors subject to legal requirements and competitive constraints;

- conserve, protect, and increase the owners/investors' assets; and

- respect owners/investors' requests, suggestions, complaints, and formal resolutions.

Suppliers

Our relationship with suppliers and subcontractors must be based on mutual respect. We therefore have a responsibility to:

- seek fairness and truthfulness in all our activities, including pricing, licensing, and rights to sell;

- ensure that our business activities are free from coercion and unnecessary litigation;

- foster long-term stability in the supplier relationship in return for value, quality, competitiveness and reliability;

- share information with suppliers and integrate them into our planning processes;

- pay suppliers on time and in accordance with agreed terms of trade; and

- seek, encourage and prefer suppliers and subcontractors whose employment practices respect human dignity.

Competitors

We believe that fair economic competition is one of the basic requirements for increasing the wealth of nations and ultimately for making possible the just distribution of goods and services. We therefore have a responsibility to:

- *foster open markets for trade and investment;*

- *promote competitive behavior that is socially and environmentally beneficial and demonstrates mutual respect among competitors;*

- *refrain from either seeking or participating in questionable payments or favors to secure competitive advantages;*

- *respect both tangible and intellectual property rights; and*

- *refuse to acquire commercial information by dishonest or unethical means, such as industrial espionage.*

Communities

We believe that as global corporate citizens we can contribute to such forces of reform and human rights as are at work in the communities in which we operate. We therefore have a responsibility in those communities to:

- *respect human rights and democratic institutions, and promote them wherever practicable;*

- *recognize government's legitimate obligation to the society at large and support public policies and practices that promote human development through harmonious relations between business and other segments of society;*

- *collaborate with those forces in the community dedicated to raising standards of health, education, workplace safety and economic well-being;*

- *promote and stimulate sustainable development and play a leading role in preserving and enhancing the physical environment and conserving the earth's resources;*

- *support peace, security, diversity and social integration;*

- *respect the integrity of local cultures; and*

- *be a good corporate citizen through charitable donations, educational and cultural contributions, and employee participation in community and civic affairs.*[3]

C.5. CERES Principles

The CERES coalition is a group of U.S. environmental, financial, and advocacy organizations with a commitment to sustainability and continuous environmental improvement. The CERES Principles consist of a 10-point code of environmental conduct. They were originally known as the Valdez Principles, developed in 1989 after the Exxon Valdez oil spill. Currently, they are endorsed by 70 companies.

PRINCIPLES

Protection of the Biosphere

We will reduce and make continual progress toward eliminating the release of any substance that may cause environmental damage to the air, water, or the earth or its inhabitants. We will safeguard all habitats affected by our operations and will protect open spaces and wilderness, while preserving biodiversity.

Sustainable Use of Natural Resources

We will make sustainable use of renewable natural resources, such as water, soils and forests. We will conserve non-renewable natural resources through efficient use and careful planning.

Reduction and Disposal of Wastes

We will reduce and where possible eliminate waste through source reduction and recycling. All waste will be handled and disposed of through safe and responsible methods.

Energy Conservation

We will conserve energy and improve the energy efficiency of our internal operations and of the goods and services we sell. We will make every effort to use environmentally safe and sustainable energy sources.

Risk Reduction

We will strive to minimize the environmental, health and safety risks to our employees and the communities in which we operate through safe technologies, facilities and operating procedures, and by being prepared for emergencies.

Safe Products and Services

We will reduce and where possible eliminate the use, manufacture or sale of products and services that cause environmental damage or health or safety hazards. We will inform our customers of the environmental impacts of our products or services and try to correct unsafe use.

Environmental Restoration

We will promptly and responsibly correct conditions we have caused that endanger health, safety or the environment. To the extent feasible, we will redress injuries we have caused to persons or damage we have caused to the environment and will restore the environment.

Informing the Public

We will inform in a timely manner everyone who may be affected by conditions caused by our company that might endanger health, safety or the environment. We will regularly seek advice and counsel through dialogue with persons in communities near our facilities. We will not take any action against employees for reporting dangerous incidents or conditions to management or to appropriate authorities.

Management Commitment

We will implement these Principles and sustain a process that ensures that the Board of Directors and Chief Executive Officer are fully informed about pertinent environmental issues and are fully responsible for environmental policy. In selecting our Board of Directors, we will consider demonstrated environmental commitment as a factor.

Audits and Reports

We will conduct an annual self-evaluation of our progress in implementing these Principles. We will support the timely creation of generally accepted environmental audit procedures. We will annually complete the CERES Report, which will be made available to the public.[4]

C.6. Enlibra

Enlibra stands for a set of principles for protecting air, land, and water, proven effective in resolving environmental and natural resource disputes. The word Enlibra was coined by the Western Governors' Association to symbolize balance and stewardship

- *National Standards, Neighborhood Solutions—Assign responsibilities at the right level*

- *Collaboration, Not Polarization—Use collaborative processes to break down barriers and find solutions*

- *Reward Results, Not Programs—Move to a performance-based system*

- *Science for Facts, Process for Priorities—Separate subjective choices from objective data gathering*

- *Markets Before Mandates—Pursue economic incentives whenever appropriate*

- *Change a Heart, Change a Nation—Environmental understanding is crucial*

- *Recognition of Benefits and Costs—Make sure all decisions affecting infrastructure, development and environment are fully informed*

- *Solutions Transcend Political Boundaries—Use appropriate geographic boundaries for environmental problems*[5]

C.7. Hannover Principles

In 1991, the city of Hannover, Germany, asked William McDonough to write out the general principles of sustainability for the 2000 World's Fair. The result was the creation of nine principles now known as the "Hannover Principles," which have become the foundation for sustainable design.

Insist on rights of humanity and nature to co-exist in a healthy, supportive, diverse and sustainable condition.

Recognize interdependence. The elements of human design interact with and depend upon the natural world, with broad and diverse implications at every scale.

Expand design considerations to recognise even distant effects. Respect relationships between spirit and matter. Consider all aspects of human settlement, including community, dwelling, industry and trade in terms of existing and evolving connections between spiritual and material consciousness.

Accept responsibility for the consequences of design decisions upon human well-being, the viability of natural systems, and their right to coexist.

Create safe objects of long-term value. Do not burden future generations with requirements for maintenance or vigilant administration of potential danger due to the careless creation of products, processes or standards.

Eliminate the concept of waste. Evaluate and optimise the full life-cycle of product and processes to approach natural systems, in which there is no waste.

Rely on natural energy flows. Human designs should, like the living world, derive their creative forces from perpetual solar income. Incorporate this energy efficiently and safely for responsible use.

Understand the limitations of design. No human creation lasts forever, and design does not solve all problems. Those who create and plan should practice humility in the face of nature. Treat nature as a model and mentor, not an inconvenience to be evaded and controlled.

Seek constant improvement by the sharing of knowledge. *Encourage direct and open communication between colleagues, patrons, manufacturers and users to link long-term sustainable considerations with ethical responsibility, and re-establish the integral relationship between natural processes and human activity.*

The Hannover Principles should be seen as a living document committed to transformation and growth in the understanding of our interdependence with nature, so that they may adapt as our knowledge of the world evolves.

C.8. Bellagio Principles[6]

In November 1996, an international group of measurement practitioners and researchers met at the Rockefeller Foundation's Study and Conference Center in Bellagio, Italy, to review progress on sustainable development. The Bellagio Principles consider the four aspects of assessing progress toward sustainable development: (1) the establishment of a vision of sustainable development and clear goals, (2) the content of the assessment and the need to merge a sense of the overall system with a practical focus on current priority issues, (3) the key issues of the process of assessment, and (4) the necessity for establishing a continuing capacity for assessment.

1. Guiding Vision and Goals
Assessment of progress toward sustainable development should be guided by a clear vision of sustainable development and goals that define that vision

2. Holistic Perspective
Assessment of progress toward sustainable development should:

- *include review of the whole system as well as its parts*

- *consider the well-being of social, ecological, and economic sub-systems, their state as well as the direction and rate of change of that state, of their component parts, and the interaction between parts*

- *consider both positive and negative consequences of human activity, in a way that reflects the costs and benefits for human and ecological systems, in monetary and non-monetary terms*

3. Essential Elements
Assessment of progress toward sustainable development should:

- *consider equity and disparity within the current population and between present and future generations, dealing with such concerns as resource use, over-consumption and poverty, human rights, and access to services, as appropriate*

- *consider the ecological conditions on which life depends*

- *consider economic development and other, non-market activities that contribute to human/ social well-being*

4. Adequate Scope
Assessment of progress toward sustainable development should:

- *adopt a time horizon long enough to capture both human and ecosystem time scales thus responding to needs of future generations as well as those current to short term decision-making*

- *define the space of study large enough to include not only local but also long distance impacts on people and ecosystems*

- *build on historic and current conditions to anticipate future conditions—where we want to go, where we could go*

5. Practical Focus

Assessment of progress toward sustainable development should be based on:

- *an explicit set of categories or an organizing framework that links vision and goals to indicators and assessment criteria*

- *a limited number of key issues for analysis*

- *a limited number of indicators or indicator combinations to provide a clearer signal of progress*

- *standardizing measurement wherever possible to permit comparison*

- *comparing indicator values to targets, reference values, ranges, thresholds, or direction of trends, as appropriate*

6. Openness

Assessment of progress toward sustainable development should:

- *make the methods and data that are used accessible to all*

- *make explicit all judgments, assumptions, and uncertainties in data and interpretations*

7. Effective Communication

Assessment of progress toward sustainable development should:

- *be designed to address the needs of the audience and set of users*

- *draw from indicators and other tools that are stimulating and serve to engage decision-makers*

- *aim, from the outset, for simplicity in structure and use of clear and plain language*

8. Broad Participation

Assessment of progress toward sustainable development should:

- *obtain broad representation of key grass-roots, professional, technical and social groups, including youth, women, and indigenous people—to ensure recognition of diverse and changing values*

- *ensure the participation of decision-makers to secure a firm link to adopted policies and resulting action*

9. Ongoing

Assessment Assessment of progress toward sustainable development should:

- *develop a capacity for repeated measurement to determine trends*

- *be iterative, adaptive, and responsive to change and uncertainty because systems are complex and change frequently*

- *adjust goals, frameworks, and indicators as new insights are gained*

- *promote development of collective learning and feedback to decision-making*

10. Institutional Capacity

Continuity of assessing progress toward sustainable development should be assured by:

- *clearly assigning responsibility and providing ongoing support in the decision-making process*

- *providing institutional capacity for data collection, maintenance, and documentation*

- *supporting development of local assessment capacity*

---

[1] The Earth Charter International Secretariat, University for Peace Campus - P.O. Box 138-6100 - San José, Costa Rica. http://www.earthcharter.org/ © William A. Wallace C-1 file: O SDfEngrs Apndx C SD PRINCIPLES Final v1a

[2] Stephen B. Young, Global Executive Director , 401 N. Robert Street, #150, Saint Paul, MN 55101

[3] Source: Caux Round Table, "Principles of Business," http://www.cauxroundtable.org/index.html.

[4] Source: CERES Coalition, "Our Work: The CERES Principles," http://www.ceres.org/our_work/principles.htm.

[5] Source: Western Governors' Association, "Policy Resolution 02-07," http://www.westgov.org/wga/policy/02/enlibra_07.pdf.

[6] Source: International Institute for Sustainable Development, Book on the Bellagio Principles, Assessing Sustainable Development: Principles in Practice, Downloadable at http://www.iisd.org/publications/publication.asp?pno=279

INDEX OF TOPICS

C

D

E

I

S

W

X

Y